Study Skills

CW00725162

"The only source of knowledge is experience."

"I have no particular talent, I am just curious."

"Imagination is more important than knowledge."

"The only thing that interferes with my learning is my education."

Albert Einstein

Study Skills Made Easy

*A Problem-Based Study Skills Guide
for Engineers and Scientists*

Roy Gregory

MechAero

MechAero Publishing
46 Lancaster Road
St Albans
AL1 4ET
UK

www.mechaero.co.uk

© MechAero Publishing 2005

The right of Roy Gregory to be identified as the author of this work has been asserted by him in accordance with the United Kingdom Copyright, Designs and Patents Act 1988

All rights reserved; no part of this publication may be reproduced, stored in any retrieval system, or transmitted in any form, or by any means, electronic, mechanical, photocopying, recording or otherwise, except as specifically authorised in accordance with the conditions of purchase, without either the prior written permission of the Publishers, or a license permitting restricted copying in the United Kingdom issued by the Copyright Licensing Agency Ltd., 90 Tottenham Court Road, London W1P 9HE

Published by MechAero Publishing, 2005, 2006
ISBN 0-954-07344-4

Set by MechAero

Printed in Great Britain by TJ International Ltd
Padstow PL28 8RW

Contents

Chapter 3

Chapter 4

Chapter 5

Preface

I have had the privilege of working in education all my life and have always been fascinated by the learning process. Human beings from an early age have the innate need to explore and change their environment and communicate with each other. I have so often been impressed by the variety of ways that humans learn and the commitment and tenacity with which so many students apply themselves to their tasks.

I found myself at an early age struggling with my own education. Eventually I was successful after finding my own ways and paths to learning. I found that the natural ability to learn needed to be rescued from an education which, by its methods and atmosphere, created a sense of fear and failure.

My engineering degree led me into lecturing, which in turn led me to research and training in the field of education. My interest in those who need alternative routes also led me to become a tutor at the Open University. This book takes from all those places and particularly from my students, from whom over the years I have learned so much.

My intention was to produce a book that did not provide prescription, but was a starting point with approaches and ideas that can be developed to suit individual needs. It has been targeted at scientists and engineers, since these are the students I have had most to do with and think I know most about. I have tried to write it in a simple, functional way and have put the problem-solving method at its heart.

I hope this book helps all who use it to enjoy learning and study all their lives as much as I have in mine.

Roy Gregory BSc MPhil MRAeS CEng MA
St Albans 2005

Acknowledgments

Having taught study skills for many years, this book is the fruit of numerous courses at undergraduate and postgraduate levels. My interest was originally awakened by David Jacques, to whom I will always be grateful for opening up such a fascinating field. Lin Thorley and Christine Shepperson are two co-workers whose influence is in all that I have written, and have always provided encouragement, advice and an enjoyment in working with them.

This book would not have been published without the help and encouragement of Ray Wilkinson, a colleague at the University of Hertfordshire, who has kept me going when I perhaps would have stopped.

Introduction

Study skills in engineering and science

Photo: NASA

When a space vehicle is launched at Cape Kennedy it represents not only the culmination of a large and complex mixture of engineering and science expertise and skills, but also a massive exercise in problem solving, team work, learning, planning and communication. These latter, more 'people-orientated' skills are often called 'core skills', 'key skills' or 'personal skills' and are highly valued by employers. They are very similar to the 'study skills' needed to be an effective student. Study skills are the subject of this book, which covers *Problem Solving*, *Managing Yourself*, *Managing Your Learning*, *Team Work*, *Communication*, *Writing Reports* and *Giving Talks*.

Developing your study skills

Much of the improvement in these skills can only be achieved by practice followed by feedback (i.e. information of how

Example

Appendix I describes an example of a problem involving many engineering, scientific and personal skills, which shows the importance of many of the skills described in this book. It involves the investigation into what became known as the 'Kegworth Air Crash'.

It was at its core a communication problem between engine and pilot concerning the shutdown of a particular malfunctioning engine. The various skills used included:

- solving the problem of why the aircraft crashed;
- managing the investigation and learning and gathering data as the project progressed;
- communicating and listening to all the people concerned, and communicating within the investigation group;
- presenting the results in talks and in reports to various agencies.

The case study shows how all the skills you will be developing through this book are essential for scientists and engineers. It is also an example of the way a technical report is written.

well you have done and what could be improved), followed by working out what to do next to improve, and followed by more practice. Sitting in a lecture hearing someone talk about how to write a report can be helpful but it must be reinforced by the practice/feedback cycle. An important skill that you will need to develop as part of this cycle is to assess your own performance, as well as to find ways of getting feedback from others (colleagues at work and at college, supervisors and tutors, family and friends). This book gives you the basic knowledge you need to improve and suggests opportunities for practice through exercises integrated in the text.

The book's subtitle suggests that each topic can be approached as a problem to be solved. Because of this, although the book can be read in any order, it will be helpful to read the *Problem Solving* chapter first. The ability to solve problems is a skill of fundamental importance for students studying science and engineering, and this will get you off to a good start. You will be encouraged to return to this problem-solving process in each chapter. Looked at in this way, you are looking for solutions to the problems of:

- How can I manage myself?
- How can I manage my learning?

- How can my team work best together?
- How can I communicate well?
- How can I write good reports?
- How can I give good talks?

You will need to consider SOLUTIONS, which will consist of ideas, approaches and techniques, and ways of changing yourself. Each chapter represents a concise guide to cover the main areas to start you off. In many cases the text is presented as bullet points, checklists or questionnaires so it is easy to read and accessible. The chapters do not represent complete solutions or methods, but ideas and suggestions which you can build on and/or modify.

From all this you will need to SELECT what is appropriate for your own needs and situation. These selected 'solutions' are then tested to see whether they solve the problem (EVALUATION). The process is one that goes on continuously, since there are always more improvements that can be made. The problem-solution cycle is shown in the figure.

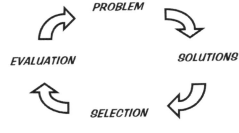

At the end of each chapter there is an exercise which asks 'How are you doing?' which suggests that you go round the cycle again to make sure you are solving the problem in the best possible way.

As a start the exercise below asks you to look at some of the basic skills covered in the book, and to make an assessment of yourself and what you can do to improve.

When you consider ways to improve, you should be as specific as possible. Choose definite and measurable actions such as going on courses, plans for practice in specific ways, new techniques to try, how feedback will be obtained with times when this will be done. The plans should be SMART (see Chapter Two). That is, they should be:

SPECIFIC
MEASURABLE
ACHIEVABLE
REALISTIC
TANGIBLE

It is no use, for example, just saying 'I will improve my time management' without planning to do things differently in a specific way. Change and development come when you do things differently.

Exercise 0.1		
Consider the list of skills listed below, and write down, for each one, your strengths and weaknesses and what you can do to improve your weaknesses. You could get some feedback from a friend, tutor or fellow student by showing them this list and asking them to comment on the accuracy of your self-assessment.		
Skill	**Strengths**	**Weaknesses**
Managing myself		
How I will improve:		
Managing my learning		
How I will improve:		
Team work		

Skill	Strengths	Weaknesses
How I will improve:		
Writing reports		
How I will improve:		
Giving talks		
How I will improve:		
Comments from a friend, etc., on the accuracy of the above:		

Personal development planning

As mentioned previously, study skills are similar to personal skills. A more-formalised process is often introduced into academic courses and into workplace appraisal to develop these. This is called Personal Development Planning (PDP). The words used are sometimes different but the basic cycle is the same.

The PDP cycle can be presented as:

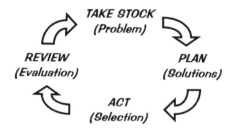

TAKE STOCK
(Problem)

REVIEW
(Evaluation)

PLAN
(Solutions)

ACT
(Selection)

If you go back to the completed Skills List you have produced in Exercise 0.1, you will see that you have *Taken Stock* in identifying your strengths and weaknesses, and in deciding how you will improve you have engaged in *Planning*. The cycle is completed by *Acting* and then by *Reviewing* (which you are asked to do in exercises at the end of each chapter when asked to respond to the question 'How are you doing?').

Starting your own PDP file

The exercises at the end of each chapter could form the basis of your own Personal Development Planning File. This could be done by copying the completed exercises and storing them in a loose folder to keep a record. The ones you could consider using are shown in the table on the next page.

Most Professional Institutions now ask that as part of the development of their accredited members a CPD (Continuing Professional Development) record is kept. This works in a similar way to a PDP record but covers all the skills, knowledge, expertise, etc. which are required to work as a professional in a particular area of science or engineering. Many Institutions have their own pro forma documents for CPD. It is worth checking with the Institutions you might already be interested in, to see how you may need to continue and develop your PDP file in the future. PDP will be a good start, as well as helping you to develop your own skills in a more-conscious manner.

Exercise	Page No	Description
0.1	4	Taking stock of skills
1.5	22	Review of *problem solving* skills Plan for change
2.6	33	Review of *managing yourself* skills Plan for change
3.11	56	Review of *managing your learning* skills Plan for change
4.7	71	Review of *team working* skills Plan for change
5.10	88	Review of *communication* skills Plan for change
6.7	109	Review of *report writing* skills Plan for change
7.7	127	Review of *giving talks* skills Plan for change

You could add to or write in other ways the list of personal skills you wish to develop and how you intend to improve them. On many academic courses, students are now asked to keep their own PDP File which must be laid out in a formal way with standard pro forma documents. The processes in this book will help you keep this file on the basis of *Take Stock* (what is the problem?), *Plan* (what are the solutions?), *Act* (selection) and *Review* (evaluation).

Exercise 0.2

If you are not at present required to keep a PDP File, start one using the work in this book. Collect together, in a loose-leaf file, copies of exercises at the end of each chapter as you do them. This can be reviewed and further plans made to make sure you continue to develop in the ways you need to.

You will probably find that you will want to modify the pro formas as you gain experience of this process and will organise the file in a way that best suits your own personal development planning. You may want to identify more specific skills to develop as part of an overall plan. For example, you could decide to practice speed reading or produce a timetable of your study to improve your time management. You might decide to review your skills every few months and record this in your file. Remember also to think about the whole of your life and the way you show and develop your personal skills. You could be demonstrating considerable skill by organising a trip abroad with friends or being secretary of a Rugby Club. It is sometimes easy to miss these and think only of university, college or work. Volunteering for such activities can be way of developing your skills.

A list of exercises you might use to develop your PDP file further is shown in the table below. They do not cover every skill in the same detail but give examples which you might be able to apply to other areas.

Exercise	Page No	Description
2.2	27	Review of lifestyle Plan for change
2.4	33	Making SMART goals
3.2	41	Review of your learning Plan for change
3.5	48	Review of 'Intelligences' Plan for change
5.9	88	Review *listening* skills
6.2	94	Self-assessment of your report writing
6.3	96	Feedback from report writing
7.2	114	Self-assessment of *giving talks* skills
7.3	118	Video a talk—Plan, Act, Review
7.4	121	Plan to improve *giving talks* skills
7.6	126	Feedback from giving a talk

Chapter 1

PROBLEM: How can I solve problems?

It was suggested that you look at problem solving first because it is fundamental to science and engineering, and the techniques can be used to tackle many of the communication and study-skills issues that arise.

You will be asked to think about this in each chapter. You will see that you can use the PROBLEM, SOLUTIONS, SELECTION and EVALUATION cycle to improve your skills.

In this chapter the techniques covered concentrate mostly on defining the problem well, coming up with solutions and ways of selecting and evaluating these solutions.

Scientists and engineers are familiar with the problem-solving process. Many famous engineering failures have occurred because the right questions were not asked at the beginning or the state of knowledge was insufficient to ask them. The wobbling of the Millennium Bridge in London and the fire at London's King's Cross underground station are examples of this. Essentially engineers and scientists are 'problem solvers', whether it is a design of a product or a system, the design of an experiment, to understand a process in nature or the investigation of an accident.

Problem solving is something that we have all done since our earliest years and is a part of our natural inclination to learn and change our environment. We all use problem-solving methods, without necessarily identifying them as such, and we can be very successful using them intuitively.

Some of the common problem-solving techniques are covered in this chapter. The engineering design process has been used as an example since this is a particular formalised process used by some designers. It has similarities to other problem-solving methods and can be used for a large range of problems inside and outside the design field.

The design process

The design process is a series of steps, starting from properly understanding the problem and specifying it in detail (see Exercise 1.2). These first steps are so important  that the following section looks at this aspect more closely. It is also vital to evaluate the final solution to see that it does actually solve the problem you started with. If your problem was to manage your time to obtain a very good degree while keeping a social life, then you need to come up with a plan to do this and then consider whether it will actually do that. You can then, from time to time during the course, go back and ask the same question (i.e. review and evaluate).

Exercise 1.1

Make a list of your aims that you want to achieve during your time as a student. The problem is how to achieve these in a way that is best for you. As a start, consider each aspect of your life and list the aims in priority order.

Exercise 1.2

Choose a communication problem and apply the Design Process to it, describing briefly each one of the six stages in the table below.

The Design Process	Communication Problem
1. Understand the problem	My communication problem is:

The Design Process	Communication Problem
2. Specify in detail	More detail:
3. Propose solutions	List the ways it could be done:
4. Evaluate the solutions	Which looks the best?
5. Re-evaluate against specification	Does it communicate what I want it to?
6. Develop chosen solution—making a plan, design or strategy	More detail on the chosen plan:

Problem definition in design

One of the most impor- tant stages of problem solving, and one that is very often done poorly, is to define the problem adequately. Only when the question is properly defined can work on the  solution begin. If the problem is fully and carefully defined, you are already most of the way towards the solution. Then, when a solution has been produced, it can be checked against the problem definition to ensure it has been fully answered.

With simple problems, it is usually relatively easy to define the problem, but it is important to ensure that you step back far enough to identify the real problem, not a particular aspect based on your assumptions about the solution. For this reason, it is essential that the problem is defined before work begins on the solution. For example, if you are tasked with design- ing a new all-terrain vehicle, you might assume that it would require at least four wheels. This assumption would, however, exclude a hovercraft or a tracked vehicle.

Although the above examples have been design-based, the principle could equally be applied to other problems. Suppose your car breaks down in heavy snow. You have other people in the car, including a small child. You may think the problem is about repairing the car so you can continue your journey and get all of you safely home. However, it may be more serious than that, and may ultimately come down to a question of sur- vival. Once you have recognised this, repairing the car is only one of a range of possible solutions, which may also include calling out the emergency services, flagging down another vehicle for help, or abandoning the car and seeking help from a nearby house.

So you can see that defining the problem, in great detail, or at least in as much detail as is required to show a complete understanding, is a vital *precursor* to taking the first steps towards a solution.

You as a problem solver

During your life, you will have already successfully used various problem-solving techniques. It can be helpful to identify these so that you can more-consciously select them when you get stuck. Techniques are often transferable from one problem to another, and knowing that you already have some problem-solving tools, with a proven track record, can add to your confidence when faced with a new problem.

Exercise 1.3

Think about a problem you have dealt with in your life and/ or work where you felt really satisfied with your solution.

- What worked well to solve this problem?
- What factors made it work well?
- How did I start?
- How did I know it was successful?
- What skills did I use?
- What techniques did I use? (Quickly look through this chapter to see the range.)
- How can I use my strengths in dealing with that situation to tackle a problem I have now?

Identify below the successful problem-solving techniques you have used.

Simple techniques

We have solved problems effectively all our lives. It can be useful to identify these before we go on to more-formalised techniques.

The *problem itself can be changed* or *viewed differently*, accepted as part of life or put aside and reviewed later. Just *making a list* of what needs to be done and then prioritising items can solve some problems. The solution to a financial problem can often be started by listing income and expenditure to see where savings can be made.

Asking questions and *finding some new information* may be sufficient. These simple techniques can be particularly useful when thinking about problems like organising your study, working in a team, managing difficult people or how to communicate.

We all approach situations with some in-built ideas learned as a result of previous experiences. Often these can be a valuable source of expertise that helps to solve the problem. In fact, one common method of problem solving is to look for parallels in previous experiences that can apply, or be adapted to apply, to a current problem.

'*Trial and error*' is a simple problem-solving technique and is probably the first one that most of us would have used as a child—it is still used a great deal and can sometimes be the fastest and the most successful. It can, however, waste a lot of time if the problem is complex and the trial is not done in a systematic fashion, making a note of what was tried and what the results were. A 'diagnostic checklist' is a more-formalised version of this method, where a logical sequence of tests is designed to identify what might be wrong with, say, a computer system or the human body undergoing a health check.

The use of simple algebraic expressions or charts to calculate various options can sometimes illuminate a problem. Computer spreadsheets can be very useful tools to help this approach. A financial problem may be simply solved by writing down estimated expenditure and seeing how this can be met by income.

Charts and graphs can be used as simple problem-solving tools, which enable you to view the problem from a different perspective. Moving the desks and workstations around on a scale plan, or using a simple computer program, can help when planning a new office layout.

Lateral thinking

There are a number of techniques to help to obtain new, creative solutions to problems. The insights gained are not achieved by logic, but by allowing your brain to 'roam' unrestricted over completely new ground  that may result in innovative but effective solutions.

Sometimes our thinking can be seen as automatically going along a railway line. Lateral thinking means making a mental 'hop' to a completely different line going in a different direction to see where it takes you. Lateral thinking can also take the form of using a solution from one area of our life and applying to another.

A few of the more-common techniques to help lateral thinking are:

Reframing

This technique will sometimes be part of others. It is used to look at the problem from different perspectives, which may shed new light on the situation and change the possible solutions. When computers were reframed to be word processors the market increased to nearly everyone in the population!

Additionally, it may be possible to see positive aspects open up from an unexpected and at first sight very negative situation. For example, losing your driving licence is a negative event but it may provide an opportunity for getting fit by doing more cycling!

Brainstorming

'Brainstorming' is normally a group technique used when several problem solvers work together as a team and express their ideas in an uninhibited and usually unstructured fashion. Group members are urged to throw any idea, however wild and apparently unrealistic, into the melting pot. It provides a safe environment for lateral thinking in its broadest sense. The emphasis in the 'brainstorm' phase is on *quantity*. A wide range of ideas can be built up  as members spark ideas off each other, follow up associations and are encouraged to build on each other's suggestions.

Although the ideas are meant to flow in an uninhibited fashion, to work really well the process itself usually needs some basic ground rules. For example, there should be a method of recording ideas and a time limit is normally set (say between five minutes and half an hour), after which the team begin to make some order out of the 'chaos' of creativity. At the start someone should set the scene by defining the problem for the group.

The essential characteristics of such a session are:

• encourage unconventional ideas
• aim for quantity of ideas
• try to build on ideas put forward
• no discussion or criticism of ideas during the session

At the end of the session, the brainstorming stops and the ideas are sifted and evaluated to extract the 'gems' that can be used to provide a feasible solution.

It is important that the group do not give up half-way through allotted times. Some good new ideas often emerge after a silence when everyone has dried up. The silence often forces a different approach or angle to be suggested and the ideas start to flow again.

You can do brainstorming on your own by writing down ideas over a period of time, while you go about your daily work. Keeping a notebook near you will enable you to write down odd thoughts and ideas as they come to your mind and that otherwise can easily be lost. Keeping this by your bed will sometimes help you capture the thoughts you have when half asleep, which can sometimes bring fresh insight to a problem.

An alternative strategy is to hold a *negative brainstorming* session where everyone has to suggest ideas to make a situation worse! Afterwards the negative ideas are looked at and the team try to see corresponding positive ideas that might be useful in actually solving the problem.

There are computer programs available, such as *Thought-Path* and *BrainStorm*, to help with this brain-storming process which you might find worth looking at to see if they suit you. The ideas are recorded on the computer and the program allows linking in a way similar to a mind map (see page 49). Microsoft OneNote, although not strictly brainstorming software, encourages non-linear thinking, which can also be helpful in this kind of environment.

Scenario writing

This is a problem-solving tool where you develop a picture of the future. The scenario should be a logical outgrowth of current or recent events and should develop and predict future outcomes. As you gather and integrate information into a future scenario it may develop into a programme for solving a current problem.

It may, for example, provide insights into the ways in which a situation might develop. The scenario may help to pinpoint critical areas that may need changing to alter an undesirable 'future', or to

What next?

discover some areas that need building on to promote a desirable 'future'. The very act of 'telling the story' may generate ideas about problems not so far foreseen or, more helpfully, trigger new ideas about solving ones already known about. It may even provide a new slant on situations currently regarded as problems and allow you to see them as potential assets instead.

Another slant to this technique is to describe the best-possible outcome and write a story, however unreal or 'unthinkable', in which the outcome actually happens. It is sometimes described as 'thinking the unthinkable'.

Means-ends analysis

This technique was developed to allow a computer to solve mathematical problems by identifying the difference between current information on a problem and the information needed to solve the problem. A reduced version of this can be used as a more-general problem-solving tool. If you have a problem where the solution is fairly obvious but the means by which that solution is to be achieved is difficult, you can start with the answer and then 'write backwards' to develop a picture (or several possible pictures) of how the answer could be reached.

For example, the answer to a production problem may be that a particular piece of equipment must be available for a month in June or July. This may be difficult because of cost, other users, lack of staff expertise on the equipment, unreliability of the equipment, etc. Starting from the 'end' (i.e. that piece of equipment is the only answer to the problem) you plan backwards towards a solution that overcomes the obstacles.

Decision aids

Sometimes the problem is to decide between a number of options, and some of the more common tools for that are given below.

Cost-benefit analysis

When choosing between several different potential solutions to a problem it can be useful to look at the benefits and the costs involved and judge whether the cost incurred is worth the benefits gained.

More commonly, the cost and benefits will be a mixture of quality issues (such as life style, environmental impact) and quantity issues such as monetary cost, or expressed as a scientific measurement, such as faults per thousand in a production line.

Decision tree

A decision tree is a diagrammatic representation of a series of choices and the paths and outcomes to which they lead. It displays the inter-relationship between the elements of the problem. Like a tree the diagram starts from a single base or trunk (the decision node) and branches at each point where a possible pathway diverges (probability nodes) until the various end points are reached at terminal nodes (at the ends of the 'twigs'). These end points show the possible outcomes or pay-offs that might result from taking that pathway or branch.

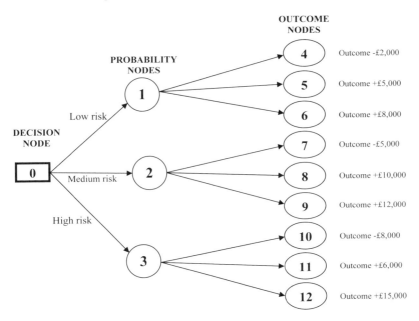

Feasibility matrix

This is a technique for deciding between a set of options, possible solutions or strategies. The attractiveness of an idea or option under consideration is set against its feasibility. Each idea or option is rated high, medium or low on each variable. The most promising are obviously those that score highly on both ratings.

An example of a feasibility matrix is given below.

Feasibility

	High	Medium	Low
High			*Option A*
Medium	*Option B*		
Low			*Option C*

Attractiveness (label on left side of table)

Exercise 1.4

Taking a course of study and changing your priorities in life has cost benefits. Produce a cost-benefit analysis of the course you are undertaking, or thinking of undertaking. For example, getting a First-Class Honours degree or captaining a University sports team will have costs and benefits which are useful to identify. One could be at the cost of the other!

How are you doing?

Exercise 1.5

From time to time it is useful to check on how well you are using all your abilities and all the techniques you know to *solve a problem*. Use the table below to consider 'How are you doing?' with a specific problem you have at present, using the questions to check this out. (It would be useful to you if this were a study-skills problem.)

QUESTIONS	NOTES AND COMMENTS
Problem What do I want to achieve? Have I defined the problem in sufficient detail? How will I know when I have achieved my aims?	
Solutions Have I considered all the solutions? Have I considered all the techniques? Have I made use of all my qualities and skills?	
Selection Have I consciously selected the best solution? Have I made a plan of what, where, when…? Does it solve the problem?	
Evaluation Have I achieved my aims (solved the problem)? Could the solution be improved?	

Chapter 2

PROBLEM: How can I manage myself?

Studying usually demands considerable sacrifice and it is important that the process is managed. This includes managing yourself and your time.

Managing yourself involves both personal qualities and techniques. This chapter covers the most important of these. It is not exhaustive but does give you a starting point. You need to SELECT and EVALUATE what works for you in your situation and go round the problem-solution cycle until you have achieved what you want to achieve.

There are many books on the market on how to manage yourself and on time management. It would be useful for you to read a few. As a start, you could look at the books on study themes which have been included in Chapter Eight. These have sections on time management and, with the material here, they will probably be enough. It is really common sense and self discipline but sometimes a few hints and tips can help. Hints and tips suggested by other students who have done well on their courses are included at the end of this chapter (page 34).

Physical organisation

This is an important area which is often missed, and yet is so obvious. Here are some suggestions to start you off:

- Find sufficient free space to work in and organise it well—it is best if you can have at least a study desk and shelf that is yours and remains undisturbed when you are not there.
- Organise the equipment you will use for your study— computer, filing system, reference books, etc.
- Ignore or remove interruptions, e.g. unplug or remove the phone.
- Have a watch or clock available when you are studying so you can pace yourself and see how much time you can study before you need a break.
- Keep a notepad handy at all times for instantly capturing good ideas—sometimes good ideas come when driving, walking, etc.
- Keep yourself fit (mentally and physically) for the times you are to study.

Exercise 2.1

Write down what you need to do to *prepare for study* using the above list as a guide and a prompt, although there may be other things that are important for you that are not on the list.

Planning for study

Students who do well plan for their study. Here are some suggestions for you to try:

- Write a list of tasks from all areas of your life (family, social, work, study); prioritise and set deadlines—use a to-do list for each day and each week.
- Exercise self discipline, for example prioritise your activity list and tick items off when done.
- Plan ahead with a chart (e.g. a Gantt chart, see page 32); remember to include holidays, Christmas, events from all parts of your life, assignments, etc.
- Stick to plans, but be flexible when necessary.
- Review progress regularly.
- Use a diary (paper or electronic) to make sure you remember important dates for handing in work, starting revision, etc.—give yourself reminders before hand-in dates.
- When doing sport or routine tasks use the time to think through study tasks.
- Identify free periods and use them well.
- Work at your studying 'little and often' and have time for reviewing.
- Think about when in the day you work best; whenever possible, choose to study at that time—for example are you a 'morning person' or an 'evening person'?
- Assess the time you will need for each task and then multiply it by something like two.
- Set a 'buffer' time for finishing jobs—aim to finish before the final deadline so you have time if things go wrong.
- Learn to say no to things (or people) that distract you.
- Set realistic goals for yourself, forward-plan for them and meet them.
- Write down your goals to help you to focus on your purpose and not get sidetracked.
- Commit yourself to finishing a task by giving yourself a public deadline, e.g. by asking a colleague to read a draft on a particular day.

Doing well for most of us doesn't just happen!

Exercise 2.2

Write down what you need to do differently in your life and lifestyle using the above list as a guide and a prompt. There may of course be other things that are important for you that are not on the list.

It can be useful to look at how many hours are available for study after you have done all the other things in the week. You have 168 hours available each week.

Example

Tim listed the times for various activities in the week and what was left for study. He came up with the following:

Activity	Hours	Hours left
		168
Sleeping	7 x 7 = 49	119
Eating and shopping	1 x 7 = 7	112
Going to lectures	5 x 6 = 30	82
Travelling	1 x 5 = 5	77
Dressing, washing, etc.	1 x 7 = 7	70
Sports	7	63
Leisure	6	57
Shopping, cleaning, etc.	7	50

Do you think that he was realistic? What would your list look like?

Exercise 2.3

Take a week in university or college and use the chart on pages 30 and 31 to note down all the activities in your week, to see what time is left for study. After you have completed it, note down anything that you have learnt that might change the way you manage yourself and your study.

Planning your time

This is a very important area and will help you to keep motivated. Losing motivation is a very common reason for failing or doing badly on a course. Some suggestions for planning your time are:

- Delegate tasks when possible.

- Keep meetings to the minimum time necessary; if you keep people standing it can encourage a shorter conversation! If you stand up yourself it can often indicate that a meeting is at an end.
- Reward yourself for study, e.g. plan treats when you meet deadlines—sweets, sleep.
- Quit when you are ahead; if you work or study too long you will create negative feelings and thoughts; take rest breaks.
- Know your own limits and needs.
- Keep yourself fit and 'work-ready'. Remember that 20 units of alcohol take 20 hours to clear the system!—hangovers on weekends don't help when you need to get a lot of work done!
- Utilise quiet times where you live, e.g. early morning, weekends.
- Make 'robust' resolutions that fit in with your lifestyle—ones that you are likely to keep.
- Explain the course to your friends and family so that they know how important it is to you and why you need to study—studying changes your lifestyle.
- Make sure you have a balance between study and leisure/social.

Diary for a week

Time

	0	1	2	3	4	5	6	7	8	9
Sunday										
Monday										
Tuesday										
Wednesday										
Thursday										
Friday										
Saturday										

(24-hour clock)

10	11	12	13	14	15	16	17	18	19	20	21	22	23

How many hours are left for study?

SMART goals

Goals can be useful in planning your study—large ones such as finishing the design work for a project or starting to write a project report, and small ones like what work you will do over a weekend. A useful way of helping to ensure that the goals are met is make sure they are as near as possible to SMART goals, i.e.:

Specific—where, when, what?

Measurable—the where, when, what should (as far as possible) be numbers or quality judgements that can be made, i.e. "how will I know when I have achieved my goal?"

Achievable—do you actually have the means, resources and time to achieve them?

Realistic—are you both willing and able?

Tangible—will you be able to experience the goal with one of your senses? See it, hear it, feel it...

Gantt chart

A Gantt chart is commonly used in science and engineering. It sets out all the activities of a project against a time line. This helps to bring reality to your plans by seeing them on paper, enables you to see where you might be overloaded, helps reorganise things and also see the consequences when things 'slip', and helps to identify what must be done to change the plans.

Below is an example of the first part of a Gantt chart for a building project to give the idea.

OPERATION	Feb	March	April	May	June	July
Design and Plans	———	——►				
Costing		—·—·—	·—·—·—	·—·—·—►		
Building Regulations and Planning Permission		———	———	——►		
Financial arrangements				—·—·—	·—·—►·►	
Builders						
Electricians						
Plumber						
Inspections						I
Roofers						
Carpenters						—·—·—
Decorators						

Exercise 2.4

Write a list of goals for some area of your life and try to make the goals as SMART as you can.

Exercise 2.5

Make a Gantt chart for your study over a reasonable period of time. Put on it all that you plan to do, what holidays you will take, other events you know will happen, examinations you will take, revision times, deadlines that you know of, etc.

Example

A group of students who had done well on their course were asked for their 'hints and tips' in a group brainstorm. This is what they came up with:

- Things take longer than you think—add at least 50% to the maximum you first think of.
- We think we will remember but we don't—write down and organise.
- A normal week or month has the unexpected in it—'plan' for that!
- We can't always study as we did at school—change strategies that now don't work.
- Sometimes you have too many things to do, so you don't start anything—start somewhere and break the work down into small, manageable pieces.
- Feeling confused can be part of learning—it can be encouraging to work in a group.
- Be strategic in your study—think about what you spend your time on and why; think about the marks you are gaining and what you are learning from what you are doing.
- Work out a time vs effectiveness graph—a break may increase your effectiveness when you come back to study (say after a 15 minute coffee break).
- Mobilise your support network and balance your life (family, friends, relationships, social, sport, interests, leisure, work, health, etc.)—think about everyone you know and how they might help your study.

How are you doing?

Exercise 2.6

Use the table below to consider 'How are you doing' to solve the problem of managing yourself. Use the questions to check this out.

QUESTIONS	NOTES AND COMMENTS
Problem What are my aims for managing myself? Have I defined them in sufficient detail? How will I know when I have achieved my aims?	
Solutions Have I considered all the solutions? Have I considered all the techniques? Have I made use of all my qualities and skills?	
Selection Have I consciously selected the best solution or plans? Have I made a plan of what, where, when...? (Is it SMART?) Does it solve the problem?	
Evaluation Have I achieved my aims (solved the problem)? Could the solution be improved? In the future I will...	

Chapter 3

PROBLEM: How can I manage my learning?

Managing your learning involves both personal qualities and techniques. This chapter covers some of the more common ones. You will of course discover others for yourself, but it does give you a starting point. You need to SELECT and EVALUATE what works for you in your situation and go round the problem-solving cycle until you have achieved what you want to achieve.

You are first asked to consider how and where you learn best and your qualities as a learner. You will already have found ways of learning that work for you and it is good to identify these. They may not all work as your circumstances change, but some will. You can improve the ones that work, identify the ones that don't and find new ones that do.

You are then given some guidance on the more basic skills that students often find difficult. There are sections on Study Techniques, Taking Notes, Mind Maps, Speed Reading, Information Retrieval, Revision, Examination Techniques and working in Self-Help Groups.

Remember that students often find that working together in small 'self-help' groups can be of great help. Try tackling a piece of work such as some tutorial questions or an essay topic together in a small group and see if this helps. If it does, it would be best to use this method from the beginning.

There are many books on the market about how to study and it would be useful if you read a few. It can be confusing to have such wide choice, so you have been given a short list of useful books in Chapter Eight.

I learn best when...

Think about how you learn and what works for you. Think back over a 'learning incident' in your life. A learning incident is an occasion when you learned something significant.

From these, try to identify statements about when you learn well and when you do not learn well. Some examples are given below, which won't be yours but might help you identify your own list.

Examples of learning incidents

- Learning to drive
- Taking up a foreign language
- Dealing with problems of a difficult relationship
- Planning a charity event
- Doing a major report and presentation for clients
- Learning how to use a new computer package
- Learning to cope with a disability
- Taking a training course at work
- Learning to take minutes of meetings

Examples of 'I learn best when...'

'I am a person who learns well when...
   ...I have clear objectives and I feel what I am learning is
      valuable to me
   ...the objectives are ones I chose and want to achieve
   ...I feel relaxed and in control of what I am learning and the
      pace at which I learn
   ...I have encouragement from others
   ...I can build on what I know already
   ...I use a variety of learning sources—books, lectures,
      manuals, one-to-one tuition, hands-on practice'

'I am a person who does not learn well when.....
   ...I feel inadequate because the pace is too fast, things are
      not well explained and I cannot keep up
   ...I am not given any help or encouragement
   ...I am expected to learn things that do not seem to be
      very useful to me'

Exercise 3.1

Write your own list of:

'I am a person who learns well when...'	
'I am a person who does not learn well when...'	
Write down what implications this list has for the way you are to plan your study.	

Learning styles

As soon as you start to think about your own learning, you come to realise that you have different approaches to learning from other people. Some prefer to start with the 'big picture' and others begin with details that gradually build up. Some people prefer to approach learning as 'doers' and others as 'watchers'.

These ideas have been developed and linked to stages of learning and learning styles.

Stages of the Learning	Learning Style
A focus on personal experience and the 'here and now' involved with feeling and intuition.	ACTIVIST
A focus on reflection after observation, involved with understanding rather than a practical application.	REFLECTOR
A focus on logic, ideas, concepts and theories.	THEORIST
A focus on practical application using real things and involving activity.	PRAGMATIST

The four styles are described below:

Activists

Learn best by constant exposure to new experiences. Readily become involved in and enthusiastic about new ideas. Often act first, think later. Enjoy a fresh challenge but get bored with detail, consolidation and tedious routine. Learn least well if they have to take a passive role.

Reflectors

Learn best when they have time to consider and assimilate experiences. Cautious and thoughtful, they consider all angles before deciding. Often listen and observe rather than participate. Learn least well when they have to take rapid action without time to plan.

Theorists

Learn best when they can integrate information into sound and logical theories. Think problems through step by step, assimilating facts into a rational pattern. They enjoy analysis and the use of models to illustrate points. Tend to be uncomfortable with personal opinions and creative, unsubstantiated thinking. Learn least well when they are not able to research in depth.

Pragmatists

Learn best when something has clear practical value that can be applied and tested. Tend to be blunt, direct and often impatient with open-ended discussion. Learn least well when learning is unrelated to a practical purpose.

Some people show a strong tendency to one or two styles, or a strong aversion to one or more. Others are more balanced between them and use all four quite often. There are no good or bad styles. If you have a dominant style it will show you where your strengths lie as a learner. It will also show you which traits you should be wary of. People who can operate to some extent in all of the styles often benefit most from a variety of learning opportunities, if they are clear about which style is the most effective for them for a given problem.

You may already have a good idea about your preferred learning style or styles, and how much you use each of the categories.

A learning styles self-assessment questionnaire can be used to help you identify your own areas of strengths and weaknesses in working through the stages of learning (see Useful Books in Chapter Eight).

Exercise 3.2

Look back at the descriptions of the four styles and take note of two strengths and two weaknesses for each style. Identify your own preferred learning styles, then think about a learning task that you expect to deal with in the near future.

Strengths and weaknesses for each style.	
My preferred learning styles are (at least two):	
How can I tackle my task so that I utilise my learning-style strengths?	

Your qualities as a learner

Exercise 3.3

Fill in the questionnaire below. Then look at what you have filled in and think about how this might affect the way you learn and what will suit you best. Consider which characteristics are helpful and which are a hindrance to you when you want to learn and improve. At the end, write down what you might do (given the information about yourself shown in your answers) to improve your effectiveness.

CHARACTERISTIC	SELF-ASSESSMENT				
	(A) Just like me	(B) A lot like me	(C) Quite like me	(D) Not much like me	(E) Not at all like me
Well organised					
Determined					
Persistent					
Impatient					
Easily distracted					
Willing to take initiative					
Careless					
Methodical					
Cautious					
Best working alone					
Good at thinking and analysing my experiences					
Enthusiastic					
Curious					
Inventive					
Anxious					
Lacking in motivation					
Disorganised					
Late with everything					
Hard-working					
Need encouragement					
Able to manage my own studies					

CHARACTERISTIC	SELF-ASSESSMENT				
	(A)	(B)	(C)	(D)	(E)
	Just like me	A lot like me	Quite like me	Not much like me	Not at all like me
Conscientious					
Easily put off by setbacks					
Like a structured environment					
Like working in a group					
Like practical tasks					
Like theoretical tasks					
Try anything once					
Motivated by a clear aim					
Generally relaxed					
Restless					
Lazy					
What I am going to do to improve my study effectiveness:					

How do you study best?

Exercise 3.4	
Fill in the answers to these questions in the table below to help you answer the question 'how do I study best?'	
When (time of day, length of time, day of the week, etc.)?	
Where (home, library, at a desk, in an arm chair, create my own learning space, etc.)?	
What external circumstances do I need (quiet, music, with other people, etc.)?	
What motivates me most (deadlines, talking to others, practical applications, etc.)?	
What stops me studying most (television, tiredness, perfectionism, etc.)?	
What can I change?	
What has changed since I studied last time that will help or hinder me?	
What have I learnt from studying last time that can help me this time?	

Study techniques

Some well-tried and tested techniques are described below. Why not try a few that are new to you?

Get yourself mentally prepared

- Remind yourself of the purpose—'why am I studying on this course?'
- Find the "interest access" such as people, problems—'can I find anything that interests me about this topic to help my motivation?'
- Consciously calm yourself—this might be quiet, music, after exercise, etc.—whatever works for you.
- Find the right environment and time for you—morning, night time, with others, etc.

Put the information in the way that suits you

- Establish (make notes of) what you already know—this helps to encourage yourself and see what you need to learn.
- Use mind maps—see page 49.
- Keep asking questions and use those questions to help you focus your study.
- Compare lecture notes with others and discuss those things that interest you or you don't understand.
- Read a variety of books and see how each approaches the topic—one approach might suit you better.
- Involve all your senses—those that for you are most receptive (hearing, seeing, feeling?):
  - Use post-it stickers (for reminders)—stick on walls and where you will look regularly.
  - Read parts of text aloud—this can sometimes help concentration.
  - Mark a book or article in colours (use a highlighter pen).
  - Use audio tapes, videos and computers.
  - Watch TV programmes about the subject or its background.
  - Create a diagram as an overview or explanation.
  - Create the 'big picture' to see where this topic fits in.
  - Write your own notes.
- Decide on:

- What you need to 'learn by heart' and test yourself on it regularly.
- Those things you must get to understand—this can take time (so pick times when you learn best for this and don't leave to the revision times if possible).
- The things to keep practising which you might be tested on.
- Use your preferred learning styles (see pages 39-41).

Read efficiently

- Make notes as you read.
- Highlight and/or underline.
- Read the interesting chapters first—sometimes reading the last chapter or paragraph can help.
- Learn to skim-read to get an overview.
- Identify key passages/references—underline key phrases.
- Read sentences, not single words.
- Practise 'speed reading' (see page 50).

Memorise KEY points

- Use memory 'flash cards'—small cards you can keep with you so you can test yourself often.
- Use memory joggers (mnemonics, rhymes, etc.).
- Write or draw the material.
- Review as frequently as you can (each day, week, month).
- Re-read your notes—this helps the recall of information when you come to revise.
- Make some mind maps that are easy to recall.

Show that you know

- Explain and discuss with others—teaching others is a most effective way of learning.
- Test yourself often.
- Make up your own questions and answer them—you could do this for each other in groups.

Using your intelligences

Knowing your strengths in terms of your abilities is useful, and an expanded idea of intelligence can also be helpful. Although most testing assesses only two categories of intelligence

(Verbal/Linguistic and Logical/Mathematical), there are a number of other equally important and valuable ones that are not often recognised or specifically developed. Taking a look at an accepted list may help you to appreciate and develop these more in yourself, and to use them more effectively.

Intelligence	Abilities	
Verbal/ Linguistic	Reading Vocabulary Formal speech Verbal debate Creative writing	Journal/diary keeping Poetry Impromptu speaking Humour/jokes Storytelling
Body/ Kinaesthetic	Body language Role playing/mime Physical gestures Drama/dance Martial arts	'Hands-on' experiences Physical exercise Making models Inventing Sports/games
Musical/ Rhythmic	Rhythmic patterns Singing Vocal sounds/tones Percussion vibrations Music composition/ creation	Environmental sounds Instrumental sounds Tonal patterns Music performance
Logic/ Mathematic	Deciphering codes Graphic organisers Number sequences Calculation	Abstract symbols/ formulas Forcing relationships Problem solving Pattern games
Visual/Spatial	Guided imagery Drawing Colour schemes Patterns/designs Painting	Active imagination Mind mapping Sculpture Pictures

Interpersonal	Division of labour
	Collaboration skills
	Receiving feedback
	Giving feedback
	Empathy
	Managing others
		Intuiting others' feelings
		Co-operative learning
		  strategies
		Person-to-person
		  communication
		Sensing others' motives
		Group projects

Intrapersonal	Mindfulness practices Focusing/concentration Thinking strategies Emotional processing Visualisation techniques	'Centring' practices Silent reflection 'Know thyself' Guided imagery

Exercise 3.5

Fill in this Intelligences Self-Assessment and write down how this might affect your approach to your studies.

Can you make better use of some of your above-average intelligences? How might this affect your approach to study? It might help you to choose options and courses, where you can make better use of those things that you excel at. For example, if you are strong on Interpersonal Intelligence you may be better off with group work, presentations and courses that require networking to obtain results.

INTELLIGENCE	SELF ASSESSMENT				
	(A) Strong	(B) Above Average	(C) Average	(D) Below Average	(E) Weak
Verbal/Linguistic					
Body/Kinaesthetic					
Musical/Rhythmic					
Logic/Mathematic					
Visual/Spatial					
Interpersonal					
Intrapersonal					

Taking notes

Taking notes or following handouts is something that you will do in many learning situations. You may be used to it and have a good technique but it is worth thinking about why you are taking notes and the way you do it.

There are a number of reasons for taking notes, including:

- It helps concentration.
- It starts the process of making the material 'your own'.
- It helps to recall the material at a later date (in the case of handouts this can be additional notes making clearer the notes you have been given).

With notes, you need to try and pick out the essential material from what is said and get that into a legible form for future reference. It is often useful to ask the question 'What concepts are being used here?' to get to the essential part of the lecture.

It is good practice (but not always practical) to re-write your notes after the lecture. This means you can check on things you don't understand and it helps recall later by going over your notes and shifting the material from short-term to long-term memory. Conventional note taking is linear with headings and subheadings going down the page. In the next section, Mind Maps are described, which offer an alternative means of presenting material in a more-graphical form.

Mind maps

Mind Maps, which were popularised by Tony Buzan (see Useful books in Chapter Eight), involve a set of linked ideas drawn in a non-linear way—more like the interconnections made by the brain.

Mind maps can also aid memory by using shapes, diagrams, colours, etc. The more these are used, the more memorable the diagram can become, so that your visual memory comes to the aid of your conceptual memory. They can also be useful, as mentioned in Chapter One, to clarify ideas when there are many complex, interlinking pieces of information.

The example below shows how a student used a mind map to re-write his mechanics notes for the purposes of revision:

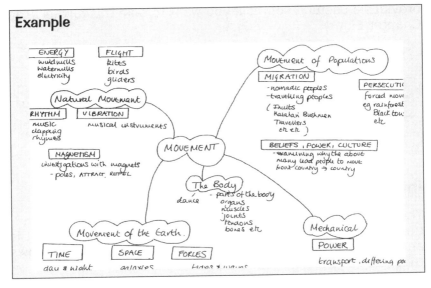

Example

Exercise 3.6

Take a topic or topics from a course you are studying and change the notes from linear format into a mind map. Did it help with understanding or remembering?

Speed reading

Most people can train themselves to learn to read more quickly than they do at present, which can have obvious advantages when studying. The technique is basically one of consciously reading lines of words and taking in the meaning as a whole rather than reading each word separately. It also helps if you have a clear idea of what you want to  get out of the text so that as you skim you can concentrate on the main ideas.

You can try this for yourself just by making a conscious effort and timing yourself. Tony Buzan's *The Speed Reading Book* provides a useful resource if you want to take this further (see Chapter Eight).

Exercise 3.7

Take a text of, say, 500 words and see how quickly you can read it to pick up the main ideas. Try to read lines and not single words. Time yourself.

Take a similar text length and see how much faster you can go. Keep practising—you will get faster.

Information retrieval

There is a wide variety of places that you can obtain information which can help you with your course and your study. It is an area where a 'brain storm' to find resources often shows that you have more available than you think.

Using the library well can make a big difference to your study. It gives you access to a large variety of books and other resources that take different approaches to the subject, and some will suit you better than others. Each discipline will have books of abstracts of conference papers and articles in journals, described briefly, so you can quickly see literature that will be useful.

Get to know your university or college library and what it has to offer, but also look outside to local libraries (in some areas you can search all of the county libraries on one data base). Think of national organisations and what they have to offer. Use book searches with companies like Amazon to give you an idea of the more popular books that are published in a particular area.

The Internet provides a wealth of useful information, but remember that the validity has not necessarily been checked by a publisher or by other academics at a conference, and must be treated with caution. You need to think about where the material has come from (does it look as though it is a reputable source?) and check it against other independent sources. Find some key websites that will give you further useful addresses to explore.

Searching in libraries and on the Internet can be overwhelming, with more information than you can cope with. Try to get an overview of the subject by 'skimming' books and articles, looking for some simple texts which give you a good base.

Using key words to search is a 'trial-and-error' problem-solving process, but the time can be reduced if you discuss it with people who know something about the subject.

It all takes time to explore and think. Remember to make notes of the references—if you don't, you can waste a lot of time looking for them again! Remember when doing a literature search to:

- write down what you know
- talk to experts first
- read conclusions of literature first as a quick way through the text
- find themes or trends
- read only the most relevant texts thoroughly
- analyse the topics for yourself
- find 'root' texts—ones that are often quoted
- read with questions in mind
- keep contents pages as a quick way to get an idea of a text

• use lists of references in books and literature to try to see the often-quoted texts.

Example

Here is the start of a 'brainstorm' of resources that might help a design project:

- People that you know—you may have experts in various fields as part of your personal network who would be pleased to help you; you may have people ready to help with your English or Mathematics, for example.
- Books—buy them, browse in libraries and bookshops, ask others which ones are helpful.
- eBooks—some are published on the Internet.
- Professional Institutions—they have libraries and journals.
- Two useful websites which give bibliographies, etc., across a wide range of subjects are BUBL Information Service (www.bubl.ac.uk) and ER-Online (www.er-online.co.uk).
- The Internet—there are many search engines, both general ones like Google and more specialist ones.
- Trade and special-interest magazines.
- National organisations publish literature, standards, books, etc.
- Libraries—they will have their own search facilities and collections of literature, and librarians can give specialist advice.
- Company publications, sales brochures, etc.

Can you think of more?

It can save you time if you search with questions in mind, such as:

- What previous work has been done?
- What basic approaches have been tried?
- What techniques have been used?
- What designs have been tried already?
- What are the basic issues?
- What difficulties have been encountered?
- What gaps in knowledge are there?
- What are the major opinions held by experts?
- What is the evidence?
- What is the authority of the texts?

Exercise 3.8

Using an Internet search engine, make a list of ten sites that can provide useful and authoritative information in your chosen field. It could be helpful to think of a topic you would like to research and see what text you can find that will quickly tell you the basics.

Revision

Revision can be tackled in many ways, and you will probably already know what works for you. The first things are to start early, and make sure the notes you will use are in good order, and if possible re-written. Any handouts should be well annotated.

Many people find it useful to précis their notes so that they can act as quick revision aids nearer the examination times (see the mind map example on page 49). Early on, you need to spend time in understanding. This can take a long time and cannot be rushed. You might need to go back to the lecturer, ask fellow students, or find other books that might take a different approach, all of which take time. The understanding will need to be complemented by 'learning by heart' some key things such as definitions and important equations. You need to work out how to do this, remembering to try using different senses. Some people learn best by hearing, some by writing and others by seeing. You can use tricks like such word association and mnemonics to help your memory (see Tony Buzan's book *Use Your Head*, Chapter Eight).

You might find you revise best on your own, although some students find that being in a group can help. A revision timetable can be useful, so that you balance all your subjects and can see what time is available and how to prioritise. It also helps to see how the many activities in your life (domestic demands, leisure pursuits, etc.) will affect the time you have for revision. What commitments have you made? What life style will maximise your chance of passing? Also, it helps you to tackle the less-enjoyable subjects that you might otherwise avoid. You want to pass them all!

Exercise 3.9	
Answer the following questions:	
'What have I found works well for me in revision?'	
'What can I try in the future that will improve my revision?'	

Examination technique

You need to approach an examination with the view that you are helping the examiners to pass you by providing all the information you can. Remember to:

- Read all the questions carefully, before deciding which to answer (if you have a choice).
- Answer the question that was set—you don't get marks by answering the question that you thought or hoped was asked.
- Plan your time so that you can answer all the questions required.
- Leave time if possible to check your answers—if you finish early, check them again.
- Remember that often the first part of a question is easier to obtain marks from than the last part—another reason for making sure you answer all the questions.
- Be as neat as you can and explain what you are doing (if you have to go to another part of the answer book to continue, say so)—it helps the examiner give you every mark s/he can.
- If you produce a ridiculous answer in a numerical question that you need to use to carry forward to the next part, tell the examiner this and that you will assume a more-sensible value to complete the calculation.
- With written questions, think in terms of what are the main points you want to get over before you start—just a minute or two planning your answer may stop you waffling and running out of time.
- If you are dyslexic or have other disabilities make sure the examiners know, so you can be given the correct support and it can be taken into account.
- Use the full time, even if is to check thoroughly what you have done—it might get you just that extra couple of marks you need for a higher grade.

Self-help groups

These are simply groups of students meeting together to help each other to study.

They are normally set up and run by students for the benefit of students. Somewhere between three and five is a good number, but pairs can sometimes work well.

Exercise 3.10	
Answer the following questions:	
'What examination techniques have I found work well for me?'	
'What can I try in the future that will improve my examination techniques?'	

Anything can be discussed which is considered of mutual benefit, and groups can be used for such things as:

- Solving tutorial questions
- Starting a project
- Sharing out work
- Helping each other with difficult topics
- Talking through work problems
- Discussing study methods, note taking, etc.
- Encouraging revision

They are a way of utilising each other's strengths for the benefit of all. The project or activity can still be your own work and presented in your own way, having gained from other students' ideas and help.

A number of things are worth considering to make it more likely that a successful group will continue.

Don't:

- Let one person do everything. Share responsibilities, e.g. convening group meetings, running meetings, taking notes.
- Run it like a class with one person 'teaching' the rest (unless someone happens to be an expert and has agreed to give a mini-lecture).
- Run it like a business meeting with a chairperson trying to get the most efficient decision.
- Use the meetings as a social occasion—humour can be helpful, but too much can make others feel it is a waste of time!

How are you doing?

Exercise 3.11

Use the table below to consider 'How are you doing?', in solving the problem of managing your learning. Use the questions to check this out.

QUESTIONS	NOTES AND COMMENTS
Problem What are my aims for managing my learning? Have I defined the problem in sufficient detail? How will I know when I have achieved my aims?	
Solutions Have I considered all the solutions? Have I considered all the techniques? Have I made use of all my qualities and skills?	
Selection Have I consciously selected the solutions? Have I made a plan of what, where, when...? Does it solve the problem?	
Evaluation Have I achieved my aims (solved the problem)? Could the solution be improved? In the future I will...	

Chapter 4

PROBLEM: How can my team work best together?

Working in groups is very common in many courses, and learning how this can work well is an important study skill.

All of us have had experiences of being part of a group or working in groups. If you watch or play team games you can see the effect that 'team spirit' has on the performance of individual players. It is as though the group has a personality of its own, which affects the individual members.

Some of the ideas presented in this chapter can help you to understand the importance of team spirit, recognise how it is developed and maintained, and find out why it can go wrong and how to put it right.

Two Video Arts films starring John Cleese (often available from college or training department libraries) are worth watching:

Meetings Bloody Meetings
More Bloody Meetings

The first video deals with the structure of a meeting, the agenda and minutes, which are also all described in this chapter. The second video deals with the more human side of the way a team works, and will help illustrate the working of some of the other ideas presented here.

Task and process

This division into task and process is one of the simplest ways of thinking about groups and the way they work.

> **Example**
>
> Terry is a fireman who regularly works in a team. They go out together to fight fires and, on occasions, join up with other fire fighters to tackle larger blazes. The team in his station spends a lot of time training together and waiting for a call.
>
> The *task* that the group must carry out is very clear. When they go out to a fire they all have specific roles. They all know what to do and who is in charge. Terry has noticed that in dangerous and stressful situations or when things start to go wrong, the *process* of their teamwork becomes increasingly important. At times like these the underlying trust that they have in each other, the way they listen and respond and their thorough knowledge of the strengths and weaknesses of all the team members are all vital.

In engineering or science, the *task* will include the more functional aspects like designing, doing an experiment, producing plans, writing minutes, obtaining resources, and defining responsibilities. The *process* will include less-tangible (but equally important) aspects like listening, communicating clearly, negotiating, co-operating and supporting.

You will see in a later model of groups (*Adair's Circles*) that the *process* is broken down into *individual* and *team* needs. In highly task-orientated situations, such as fighting fires and military operations, the needs of the team often need to take precedence over the needs of the individual. The process in Terry's case is related to the group, and maybe after the event the individual needs might need to be addressed. In military situations the group needs are addressed by the symbolism of Regimental flags and banners, and the discipline instilled to work for the group, even in situations of personal danger.

Exercise 4.1

Can you think of a situation (in your experience) where a group was so PROCESS focused they forgot about the TASK? Write a description of this situation.

Can you also think of a situation (in your experience) where a group was so TASK focused they forgot about the PROCESS? Write a description of this situation.

Now, thinking about your own experience of teams...

Exercise 4.2	
Think of a group you are in and try to identify the TASK and the PROCESS and where the emphasis is. Write the answers to the following questions:	
Would it be improved by a better balance?	
If so, how would that better balance be achieved?	

Task-group-individual balance (Adair's circles)

John Adair formulated the model that group working was made up of three elements—the task, the individual and the group. For a group to achieve the most effective results these should be in balance. He looked at each element and defined the necessary characteristics to optimise the way the group worked:

TASK—clear tasks, roles and procedures

INDIVIDUAL—clear two-way communication, a sense of individual purpose, being heard and appreciated, and fairness

GROUP—a good 'spirit', clear and agreed norms and standards of behaviour

Adair regarded these as overlapping circles, since the needs of each element would affect the others.

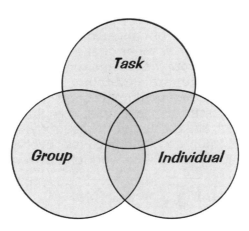

Exercise 4.3

Think of a group you are in (or have been in) and try to identify the way the Adair Circles are addressed (or not addressed). Describe what happened and evaluate the team work in terms of Adair's Circles. Write down answers to the following questions:

Were the objectives achieved (or are they likely to be)?	
Would the performance be improved by a better balance?	
How would that better balance be achieved?	

Different types of task would change the balance required. For example, situations of danger may require discipline and rapid decisions. This may mean that individual needs are less considered but the group needs and the task needs (achieving a dangerous objective) are elevated.

Some needs can be ignored in the short term, but eventually have to be addressed. For example, if a team member feels that they are not being listened to they may start to feel isolated from the group and its objectives. Eventually this must be addressed if the group is to work effectively.

Forming, storming...

Another researcher formulated a now-famous way of understanding how a new group might go through certain stages of development. The group members can be new to each other, or they can be known to each other but are in a new situation and/or a new task.

FORMING—exploring what the group will be like, finding the basis for forming relationships with others. Finding out who the other group members are and accepting whoever is in the formal or informal leadership role.

STORMING—conflicts break out as subgroups emerge, differences are confronted, control becomes an open issue and is resisted, regardless of its source.

NORMING—rules start to emerge about acceptable ways of behaving and carrying out the group task; these rules are applied to work through the areas of conflict and a spirit of co-operation develops; leadership issues are resolved.

PERFORMING—conflicts are resolved, energy is put into task accomplishment; the group is becoming effective.

> ### Example
>
> Ben was part of a newly formed group drawn from various areas in his company. After a week or so he noticed that several people in the team had different opinions about the format and importance of the progress reports presented at the regular review meetings. There were underlying tensions that were not talked about, but were making the meetings difficult and unproductive. Some people were not listening to the content of the report, but were concentrating on their own dissatisfaction with the way that it was written, and real issues were being missed.
>
> He had read the theory of group processes (above) and thought that the group needed to confront this problem (*storming*) in order to work effectively (*performing*). He raised this issue and was partially successful in getting a better understanding, but still felt that more could have been done.
>
> Next time he was part of a similar interdisciplinary team, he would suggest early on that they spent some time discussing the way they would report back and come to some agreement about the content, length and detail. He would then be able to see whether this would improve the effectiveness of the group. He had certainly learnt that these *group norms* can be important for the working of a group.

Belbin's team roles

Meredith Belbin's work on the roles that are required for a successful team has been widely used. He put forward the idea that most people prefer one or two specific roles in a group and believed that a group needs representatives of all the roles to work effectively on a task. If a dominant member takes a particular role, other members, who may have this role at the top of their 'preferred list', may have to take roles further down their list of preferences. A self-perception inventory is often used to establish an individual's preferred role(s). Belbin's roles are set out below.

BELBIN TEAM ROLES	
ROLE	**DESCRIPTION**
PLANT	Creative, imaginative, unorthodox. Solves difficult problems. Tends to ignore details and can sometimes be too pre-occupied to communicate effectively.
RESOURCE INVESTIGATOR	Extrovert, enthusiastic and communicative. Explores opportunities and develops contacts. Can be over-optimistic and can lose interest once initial enthusiasm has passed.
CO-ORDINATOR	A good chairperson. Mature and confident. Clarifies goals, promotes decision making and delegates well. Can sometimes be seen as manipulative and lacking in specific expertise.
SHAPER	Challenging, dynamic, and thrives on pressure. Has the drive and courage to overcome obstacles but can provoke others. Can hurt people's feelings.
MONITOR EVALUATOR	Sober, strategic and discerning. Sees all options and judges accurately. Can sometimes lack drive and ability to inspire others. Can be overly critical.
TEAMWORKER	Co-operative, mild, perceptive and diplomatic. Listens, builds, averts friction, and calms the waters. Can be indecisive in crunch situations and can be easily influenced.
IMPLEMENTER	Disciplined, reliable, conservative and efficient. Turns ideas into practical actions. Can be somewhat inflexible and slow to respond to new possibilities.
COMPLETER-FINISHER	Painstaking, conscientious and anxious. Searches out errors and omissions. Delivers on time. Reluctant to delegate. Can be a nit-picker.

Belbin R. M. (1981), *Management Teams: Why they succeed or fail.* Heinemann Professional Publishing, Oxford.

Example

A group on a management-training course realised that they had performed poorly in the previous exercise. They sat down to discuss what had gone wrong. When they iden-tified their likely Belbin Roles and the way this might have affected their performance, they realised that they were missing a *completer* in their team, but had a large number of *plants* and *resource investigators*. This analysis fitted in with the fact that they had performed badly mainly due to poor time management and had spent too long generating ideas.

   With this in mind they decided their performance could be significantly improved if they made a special effort to keep to time, and appointed a member of the group to be a timekeeper and progress chaser. This analysis of their per-formance, using a simple theory of the way groups work, helped them to develop an action plan for the future.

Exercise 4.4

It can be useful to think about how these roles are being fulfilled in a particular group. As you read through the roles, try to equate them to a group you know well. Consider which roles you prefer and whether these are the ones you are performing in the group. If you are not performing your preferred roles, do you know why? Is this causing any con-flict for you or other group members?

How to conduct meetings

A formal meeting is a major forum for communication in groups, and is often one where the standard of communica-tion is less than ideal. Good communications can be achieved by having rules and giving everybody an opportunity to speak. Meetings seem to have a 'bad press', because the rules are often not adhered to and the communication becomes disor-ganised (see the videos *Meetings, Bloody Meetings* and *More Bloody Meetings*).

Exercise 4.5

Think about meetings you have attended. You will probably have experienced some that were frustrating and disorganised. How many of the following problems have you seen? (Many of these are illustrated in the video *Meetings Bloody Meetings* mentioned above.) Rate the frequency of the problems you have seen as 'never,' 'sometimes', 'often' or 'nearly always'.

Problem	Frequency
Little or no preparation.	
People talking over each other and not letting others have a fair share of the time.	
No clarity between identifying the problem and discussing solutions.	
Rejections of ideas without due consideration or any attempt to build a solution from a number of suggestions.	
No clear objectives to the meeting.	
No clear recording of decisions and no indication of who will carry out actions, and by when.	
People using the occasion to impose their ideas and personality on others.	
People getting frustrated and 'switching off'.	
Can you add any more reasons to the above list? Can you see how these could be overcome?	

Communications play an important part even before the meeting convenes. The *Agenda* is an important written communication, which is sent  out in advance and can help or hinder the communications at the meeting.

Meetings in business, science and engineering tend to be of a similar type. It is important that each person in the meeting understands:

- the purpose of the meeting
- the rules by which it will operate
- how discussion/decisions are recorded and implemented (use of minutes and the role of the secretary)
- that meetings involve communication that is both listening and talking
- who is controlling the meeting (generally, in business meeting, a chairperson)

Meetings can have many purposes. To confuse or mix these up can waste a lot of time and be ineffective. Common purposes are:

- information giving (possibly with questions)
- information gathering (including 'focus groups')
- organisation (or planning)
- progress monitoring
- advisory (opinion seeking)
- persuading (presenting a case)
- motivational
- problem solving (including initial design)

It can be useful to set a time limit to the meeting. If the meeting is to be effective it must be ensured that satisfactory progress is made and that the meeting is wound up to a proper outcome. The alternative is that time is wasted, then the meeting is truncated, with no proper result and some very frustrated participants!

Exercise 4.6

Meetings often form an important part of group work on a course, and therefore it is an important to have a clear understanding of the roles of the chairperson and secretary, the function of agendas and minutes, and the way meetings are conducted. In view of their importance, more detail and examples are given in Appendix II.

Before reading the appendix, fill in the worksheet below and then compare it with the appendix. You could also ask some colleagues what they think, which might start a discussion on the way to improve the group you are in.

The role of the chairperson is...

The role of the secretary is...

The function of the agenda is...

The function of the minutes is...

Important aspects of a way a meeting is conducted are...

How are you doing?

Exercise 4.7

Use the table on the next page to consider 'How are you doing?' in a particular group, to solve the problem of working well as a group. Use the questions to check this out.

QUESTIONS	NOTES AND COMMENTS
Problem What are my aims for my group? Have I defined the problem in sufficient detail? How will I know when I have achieved my aims?	
Solutions Have I considered all the solutions? Have I considered various ideas about groups? Have I made use of all my qualities and skills?	
Selection Have I consciously selected the solution? Have I made a plan of what, where, when...? Does it solve the problem?	
Evaluation Have I achieved by aims (solved the problem)? Could the solution be improved? To improve in the future we could...	

Chapter 5

PROBLEM: How can I communicate well?

Communication is an important skill for engineers and scientists. It could be explaining something to members of your design team or presenting at a board meeting or international conference. Professional scientists and engineers need to be proficient in both oral and written communications. Communication problems are one of the major reasons for accidents such as the Hillsborough football stadium disaster, the bow doors of the Herald of Free Enterprise car ferry not being closed when sailing, or shutting down the wrong engine which caused the Kegworth air crash (see page 2 and Appendix I).

Optical illusions are an example of a communication where things are perceived to be different from reality, or where different people see things in different ways. This is a common reason for mis-communication.

Good communication can work intuitively without planning, although it does not happen especially often.

The various aspects of communication can be thought about under five headings—Purpose, Language, Structure, Delivery and Content.

They are not of course 'water-tight' commandments and they all overlap, but help to give an overview. You can use these headings as a simple check-list to help you see, when looking at your commu-nication, whether you have looked at all five aspects. In addition, this chapter includes ideas on some of the vital skills in any communication process:

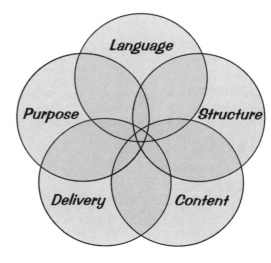

* Explaining
* Questioning
* Listening

Aspects of communication

Purpose

It would seem that an underlying purpose of any commu-nication is to achieve under-standing. This is a simplification, since com-munications often have pur-poses that are multifaceted. For example, the purpose of a television advertisement is not only to

ensure that the receiver understands something, but also to persuade them to do something (i.e. buy the product). It is important that you are clear about the purpose or purposes of your communications and identify what you are trying to achieve. Are you:

- Giving information?
- Seeking information?
- Presenting a case?
- Selling?
- Creating understanding?
- Making an appeal to people's emotions?
- Clarifying?
- Encouraging?
- Motivating?
- Persuading?
- Starting a discussion?
- Teaching?
- Explaining?

Exercise 5.1

What are the most common purposes for an engineer or scientist to produce written communications as part of his/her professional life? Write down the most common five from the above list, or choose others.

As well as the *purpose,* you need to know who your *target* audience will be, and be able answer such questions as:

- What do they need to know?
- What is their current level of knowledge?
- What will they do with the information?
- What do they need to retain?
- What is their level of motivation?
- How will they remember what they need to know?
- What is the accepted culture?
- What is our common language?
- What concepts are shared?
- What can be left out?
- What is the best medium for the content?
- What is the easiest order to present it?

Language

During your life-time you have learnt a number of languages or codes. At this moment you are

using the symbols on this page. They have been coded into symbols of written English. You are using your knowledge of that language and its written symbols to read the text and extract the meaning. You are decoding and, hopefully, under-standing the content.

Some codes are formal, such as the English with its alpha-bet, spelling, punctuation, etc. Others such as body language are more informal and involve hand movements, facial expres-sions, eye contact, nods, etc.

You could think of this as if we are *coding* and *decoding infor-mation* to attempt to convey or extract meaning, by using:

- words, symbols, numbers, pictures, graphs, charts, diagrams
- our understanding of a mutual language (including technical language, jargon, mathematics, graphical pres-entations)
- previous experiences
- body language
- assumptions
- concepts
- the structure of the communication

- our knowledge of the person sending or receiving the message
- our expectation of what the message is likely to contain, etc.

Exercise 5.2

Write a brief description of when not understanding the language has produced embarrassing or important mis-understandings in a particular situation. This could be, for example, something from your own experience, something in the media, or something historical.

This could be from a different national language or the language of a different discipline.

Structure

Structure plays such an important part in effective communication that, although we could consider it under the heading of 'coding' and 'decoding', it is so fundamental that we are discussing it

separately here. Of all the areas that can cause poor communication, structure is one of the most common and often the one that is thought about last. The term 'structure' includes the way that elements of a communication are set out. For example, diagrams, charts, photographs, maps, designs, text, symbols and numerical calculations can all form part of the structure.

The logical sequencing of the content is also part of the way the communication is structured. In a mainly text document, the order in which points are made and the way one point leads on to another are crucial structural elements.

If we expect someone to understand what we are telling them, we rarely start a story at the end, move to the middle, finish with the beginning and, somewhere in between, start a new story altogether. Nevertheless, it is surprising how often students, who have had a lot of practice in everyday communicating, fail to convey their meaning because their communication is poorly structured.

A well thought-out structure makes it easier for the receiver to understand. This is true for oral and written communication. Different professions may have preferred structures and companies may have their own ways of putting reports and presentations together. If you do not have to follow a prescribed format, a simple 'common-sense' structure will usually be the clearest and most effective way of describing a situation in a task-centred activity. One example of a simple, effective structure is to outline the situation, explain the problem, suggest a solution and evaluate the outcome. This is summarised below:

> SITUATION
> PROBLEM
> SOLUTION
> EVALUATION

This simple structure may need to be modified to fit specific occasions, but it is surprising how well a straightforward format can clarify many types of communication.

Meetings are a special, and very common, form of communication in the workplace and are described under Team Work in Chapter Four (pages 67-71).

In short communications, such as memos, the structure is obvious and does not need to be spelt out, although a logical sequence is necessary. In longer oral or written communications a clear structure is always important. A table of contents is usually worthwhile and 'sign posts' that link or provide cross-references between different sections are helpful.

An abstract at the beginning of a report can allow the reader to gain a quick overall picture of what is being discussed, which in some cases is all they need to know. The example below illustrates this:

> **Example**
>
> *"This report looks at the number of patients who come to the surgery for information, compared with those who wish to see the doctor. Several of ways providing this information, without the need for surgery visit, are discussed, proposed and costed. After examining all available data, including the results of a patient questionnaire, the report concludes by recommending that a computer-based information line should be set up, and outlines a plan for its implementation."*

In an oral presentation dealing with customer complaints, you might move from one section to the next by using a 'sign-post':

> **Example**
>
> *"We have considered all the suggestions made by our customers during our survey and have analysed them under a number of headings. We now move on to look at the viability of these suggestions and the ones we want to investigate further."*

Such linking passages are important for all forms of communication. They indicate and remind the receiver(s) of the structure and where they are in it. An old saying, which should not be taken too literally but is nevertheless useful to bear in mind, is:

> I tell them what I'm going to tell them,
> I tell them,
> and then I tell them what I told them.

> **Exercise 5.3**
>
> Write a brief description of when a poor *structure* has produced poor communication. This could be for example something from your own experience, something in the media, or something historical.

Delivery

The content will be delivered by a known code or language and by a particular means, for example:

- Spoken language
- Written language
- Sign language
- Body language
- Mathematical language
- Computer language
- Tactile language (such as Braille)

It is generally considered that more is communicated if more than one of the senses is engaged in receiving, e.g. seeing and hearing in an oral presentation.

Today there are many delivery methods available, from a simple 'talking' to a multi-media computer presentation. Visual images are readily available and can add considerably to good communication. Sometimes selecting the most suitable method is an art in itself.

Even if you are not able to choose your delivery mode, there are often opportunities to make flexible use of a single method. An overhead projector, for example, can display text, charts, maps, graphs etc. If the delivery mode is restricted to a written document, the structure, layout and choice of words can make a big difference.

One of the difficulties inherent in written communication is that you cannot be sure that you will receive any feedback from the recipient, and if you do it will be some time later. The communication needs to stand on its own and hence needs care when you produce it. The structure, grammar and spelling need to be accurate to minimise misunderstandings. (In a conversation, or even a presentation, immediate two-way interaction is usually possible and difficulties can be pinpointed and resolved at the time.) Asking someone to read a 'first draft' is a way of receiving some feedback and will help to make the final product clearer.

There are other forms of communication where the feedback is more limited, such as the telephone where you lose all body language and rely on tone of voice, pace of delivery, etc. Email is a written format, which can produce quick feedback.

Exercise 5.4

Write a brief description of when a poor *delivery* has produced poor communication. This could be for example something from your own experience, something in the media, or something historical.

Content

In thinking about the content, a number of questions need to be asked:

- What is our common language?
- What concepts are shared?
- What can be left out?
- Why am I telling/ showing them this?
- What is the best medium for the content?
- What is the easiest order to present it?

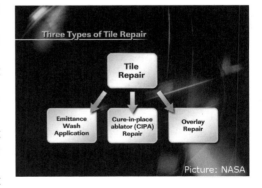

The amount to be LEFT OUT is a crucial decision since presenting more material than is necessary may just bring confusion.

Exercise 5.5

Look at a short article in a popular engineering or science magazine, and consider how well it has taken account of the purpose, language, structure, delivery and content.

Comment on these and suggest ways in which the article could be improved.

Purpose:	
Language:	
Structure:	
Delivery:	
Content:	

Communication checklist

Exercise 5.6

Use the checklist below when you are planning a communication, to check your planning process.

Have you thought about:	✓
What is the purpose of the communication?	
Who is the target audience?	
What language(s) will the audience understand?	
What structure would be appropriate?	
What is the best method of delivery (written, spoken. visual, etc.)?	
What content must be communicated?	
What interference with good communication might occur and how it can be overcome?	
What feedback can be expected?	

Explaining

This is an important skill when presenting a topic. Explaining is something we do every day and is a skill we learn with some degree of competence from an early age. Explaining has been described as:

Giving understanding to someone else

The table below gives a number of tips or guidelines that will help you to improve your 'explaining skill'. Beside each guideline is a simple example. Imagine you are explaining to a friend how to get to a restaurant situated in another town. The town is quite large and unknown to your friend, so the instructions need to be quite specific.

Guideline	Example
'Orientate' the receiver of the explanation	"You will find it most easily if you come off the motorway at Junction 5 and enter the town from the south."
Give a clear overview of what you want to communicate	"The restaurant is in the north-east corner of the town. You can reach that district by turning left off the main street at the railway station."
Develop one idea at a time	"First I will tell you where to park, and then how to get to the restaurant from there."
Obtain feedback when you can	"Are you clear about where to park, before we go on to where the restaurant is?"
Use appropriate repetition and revision	"I will just take you through that right turn again because it's a bit tricky."
Use picture or analogies	"The town's roads radiate out from the market square at the centre like the spokes of a wheel and the car park is on the spoke that goes north."
Use words and ideas that you both understand	"Do understand the directions I have given you so far?" "Are you OK with the directions at the double roundabout by the King's Arms?"
Take care to eliminate ambiguities	"There are two pubs after the crossroads on Railway Street. Make sure you turn left at the second one."
Use whatever variety of communication methods you think are practicable	"I will draw you a map as I describe it."
Concentrate on being clear and concise	"Get in the middle lane at the roundabout and that will automatically keep you on the route you need when the road forks."

Exercise 5.7

Think about how you would explain an engineering or scientific process concept such as fatigue, a chemical reaction, a black hole or how a computer works. Find a friend who knows little about the topic and use the principles listed above to explain to them.

Then think about how you might improve your skills of explaining.

Questioning

Questioning and listening are important skills in the communication process. They are particularly important where you are trying to:

- understand and obtain information (formal or informal interviews, for example);
- get feedback;
- understand or clarify what you have received.

Choosing suitable questions

The 'craft of questioning' is an important skill and it is worth listening to someone who is good at it to pick up some tips. Some extended interviews on TV can show the way the interviewer uses questions.

Here we introduce the basic types of questions you can use and the situations in which they are most helpful. Questions should be tailored to the situation and the sort of information you want. Some questions are particularly useful in specific situations; others less so.

Types of question

Open questions are worded to invite an answer with fairly detailed information. The response often includes personal opinions.

Closed questions tend to receive a very short or single-word answer.

For example, if you want to know how a scientist in your team got on at a meeting you could ask a closed question:

Example

Question: 'How did you get on at the meeting?'
Answer: 'Fine.'

This might be all you need to know. On other occasions this may not be enough. You could then ask a more open question:

Example

Question: 'Tell me about the meeting and how you got on?'
Answer: 'It was quite short and there weren't many people there. The Chairperson took the items quite slowly and summarised the decisions so I could get them down. I didn't have any serious problems.'

This might be enough information. If you wanted to learn more you might decide to use a *probe* question, such as:

Example

Question: 'Before the meeting you were worried that you might not be able to persuade them to accept your design modification. How did you to manage to do that?'

You now have much more information about how your colleague got on. The probe has particularly focused on areas where you knew there might be a problem, to see whether this had just been 'glossed over' in the first reply.

Exercise 5.8

Listen to an interview on the radio or watch one on television and pay particular attention to the questions. Try to look at the different types of questions asked and how successful they were. How would you rate the overall interview?

Types of questions:	
Rating:	

Listening

Asking the best questions that suit a situation is only one half of the equation. Even the most excellent questions are no good unless someone is listening to the answers. An effective listener 'captures' the information being sent and also picks up other signals, such as tone of voice and body language, and feeds back signals to encourage the questioner and confirm that the message is being received.

Taking notes and using audio or video recordings will help recall situations, but these are complementary to good listening at the time, not a substitute for it. Good listeners often give feedback that shows they understand, ask follow-up or clarifying questions or summarise what they have just been told as a way of confirming they have got the right message. You need to attend to:

• what is being said
• how it is being said
• what is not being said
• what feelings and emotions are being expressed, or not

Sometimes you even need to check out the words that you think you hear. This notice appeared on an office notice board:

> I KNOW YOU BELIEVE YOU UNDERSTAND WHAT YOU THINK I SAID, BUT I AM NOT SURE YOU REALISE THAT WHAT YOU HEARD IS NOT WHAT I MEANT!!!!

Exercise 5.9

Practice listening to someone or find someone to interview about their work. Identify what parts of the listening skill you found easy, and what parts hard. Try feeding back what you thought the person said and ask them to comment. Did you ask the best questions for this solution?

How are you doing?

Exercise 5.10

Take a specific communication you have had (or are having). Use the table below to consider 'How are you doing?', using the questions to check this out.

QUESTIONS	NOTES AND COMMENTS
Problem What are the aims of the communication? Have I defined the problem in sufficient detail? How will I know when I have achieved my aims?	
Solutions Have I considered all the solutions? Have I considered all the techniques? Have I made use of all my qualities and skills?	
Selection Have I consciously selected the solution? Have I made a plan of what, where, when...? Does it solve the problem?	
Evaluation Have I achieved my aims (solved the problem)? Could the solution be improved? In the future I will...	

Chapter 6

PROBLEM: How can I write good reports?

The general introduction to communication was given in Chapter Five, and this can be read as an introduction to this chapter. In particular, consider the *Purpose*, *Language*, *Structure*, *Content* and *Delivery* sections and the way they might be applied to writing reports.

A report can take many forms, from a memo comprising half a page of A4 to a large document describing months of work or research. It could be for your line-manger, for the Board of Directors, or for assessment as part of a University course.

Writing reports is an important part of the role of an engineer or scientist. Use the problem-solving model to tackle the 'problem' of writing a report. The problem involves what you want to communicate, the tools available (the words you use, diagrams, appendices, etc.), how to use the tools most effectively and checking that you have achieved the aims. When writing a report, it is essential to identify the aims of the communication—this keeps the report focused. As with any problem, the preparation is most important and you can use the headings of *Purpose*, *Language*, *Structure* and *Content* to do a lot of the work before you start typing. A mind map (see page 49) is a useful way of visualising the content and will help to put it together in a logical way.

Reports can be seen as putting together a jigsaw puzzle, to make a picture that is understood and communicates meaning.

The mode of *Delivery* is the written word. You cannot be sure that you will get feedback for written work, although in some cases it will be part of the normal process. For example, a report for a journal may be subject to editorial comments or peer review; a report prepared for a college assignment may receive detailed feedback from your tutor. If you are producing a report at work, which is going to form part of a larger document, you may get feedback from your line manager. Reports prepared for clients may produce feedback, and published reports sometimes get media coverage and/or feedback from a wider public readership.

In many cases, however, small in-house reports, routine progress reports and so on may get little response. You might then get feedback by making sure that you complete early enough to give a draft to someone for constructive criticism. When you write a report you will not necessarily be present when the content is received, and you often have less control over who reads your work than you do in other forms of communication. This means that you need to be even more careful with your language and the selection of your content.

You may find it helpful to read the section on 'Explaining' in Chapter Five. Although there are a few minor differences between oral and written explanations, many of the issues are similar.

In a report you have little control over the reader's mood when they pick up the report, but you can ensure that it is clear and well put together so that reading it is as pleasant an experience as possible! You may be able to focus the reader on what to expect by clearly stating why you are writing the report and what you expect them to get out of it. In longer reports, an Abstract or Introduction can be used to focus the reader quickly onto the important features.

Example

Karen has been asked to produce a report presenting the sales figures for all the 'reps' in her team at a Divisional Review Meeting.

As it will be a short report she will not need an Abstract but should start with an Introduction. She can use this to remind the readers why she is writing the report, to outline what she is going to write about and to guide them through the structure she has used.

The Divisional Managers and the Sales Director will be her target audience. Although they already know the background, she might sketch it in briefly because her report may be circulated more widely. Readers who were not aware of the background might draw very different conclusions if they were not given this information. She will also include details about the special circumstances in her area under 'Background', as her managers do not already know about these matters.

The structure of her report will be a conventional one (see pages 98 and 99).

She is fairly confident that she and her readers share a common vocabulary. However, as she comes from an engineering background she knows that she must aware of a tendency to assume some knowledge in her readers that they may not actually have. She will be especially careful of the words she uses in that context, and ensure that her methods of presenting data are clearly understood by managers with other backgrounds.

She will obtain feedback by asking a colleague to read her draft and to look particularly at the clarity of the statistical charts, graphs and tables.

Exercise 6.1

A good way of improving your writing is to look for good examples in professional periodicals and technical reports that you come across. Start to become aware of why articles you read (of any sort) are easy or difficult to read.

Take two reasonable-length articles (say 1000 to 2000 words), one you consider good and the other not so good. Make notes on what makes them good or not so good. In particular, look at the purpose, language, structure, content and delivery.

Essays

In many engineering and science courses, most of the written work is in the form of reports, but there will also be essays to write. This essay work usually grows in importance and frequency at post-graduate level. You are then often asked to write about topics where the structure is less obvious, although you will still need to consider all the aspects covered in this chapter. For example, you might be asked to write an essay on 'The Evidence of Climate Change', which would require a review and critical evaluation of the literature, coming to a conclusion and possibly sections on issues that are unresolved or need further work.

Essays can often still be structured using a problem-solution model (see page 79), but you might need to modify it. In most cases, applying the report-writing principles described in this chapter will still help you to write good essays.

Common mistakes checklist

When students prepare written work in science and engineering courses, there are a number of mistakes that occur very frequently.

Exercise 6.2

This is a collation of common mistakes that students make in preparing technical reports. Look at the checklist below and think about occasions when you have prepared a report or lengthy document. You may recognise these mistakes as ones you tend to make. Use the list to assess yourself against each mistake. Decide, on a scale of 0 to 10, how likely you are to make them. (0 = 'I always make that mistake' and 10 = 'I never make that mistake').

Mistakes	Your score out of 10
Mistakes in spelling, grammar and punctuation	
Sentences too long and complex	
Missing out the introduction and/or conclusion	
Not including a contents page, index, list of sources, acknowledgements section (some of these are not needed in short, informal, in-house reports)	

Poor layout, which makes reading difficult	
Diagrams, charts, paragraphs referred to in the text are hard to find	
Diagrams, graphs, charts, etc., too small, poorly labelled or have no heading	
Sequence of ideas not logical	
No clear structure, or not guiding the reader through that structure	
Disorganised material	
Not making clear the relevance of each section to the overall purpose	
Assuming too much knowledge or understanding of words from your readers	
Units of quantities not quoted	
Numbers quoted to inappropriate precision, e.g. 6.0003	
Over-ambitious curve fitting, e.g. a straight line where the points are random!	
Environmental data missed out when important (room temperature, atmospheric pressure, etc.)	
Graphs inappropriately scaled	
Figures not referred to	
Poor presentation (mistakes, untidy, poor typing)	
No attempt to quantify experimental errors	
Use of expressions like 'efficiency' or 'experimental error' with no clear definition	
No discussion of what the results mean	
Not understanding the difference between discussion and conclusions	
Conclusions not linked to aims	
No comments on whether values are as expected and, if not, why not	
No references given when needed, or bibliography given as references	
Communication not concise and complete	
Statements without justifications (don't bullshit!)	
Use of 'I' rather than third person	

Exercise 6.3

This exercise may give you a better idea about the strengths you already have, and where to improve. Identifying mistakes may also help you to avoid them.

Copy the checklist on the previous page and ask a colleague to assess a piece of your work under these headings. Compare this with your own assessment. If you think about the areas where the two assessments differ or discuss the comparison with your colleague, this will help you to focus on areas for improvement.

Structure for report writing

The structure commonly used for a formal report is shown in the table on pages 98 and 99. As you will see, it follows a similar pattern to the situation-problem-solution-evaluation described in Chapter Five.

The situation and problem will often be found together in the Introduction and Background sections. Formal reports also have sections for the Abstract, References, Bibliography and Appendices, which are extra. You should give special attention to the Abstract, Discussion and Conclusion as these sections are often done badly, or the wrong material is included under these headings.

The table follows a questioning format in which the report is written as if in answer to a series of questions. A very short report may deal with a combination of the questions under a single point, or even leave some out as inappropriate. A long report may include most of them. The way the questions are worded in the example is not rigid, and can be adapted to circumstances.

Remember, you are 'telling a story' that needs a beginning, a middle and an end, even if you are writing a thesis for a PhD.

NOTE, in the structure on pages 98 and 99, the difference between *Discussion* and *Conclusions*, and the difference between a *List of References* and a *Bibliography*. In view of the importance of these differences and the confusion that students often show about them, they are descried more fully below.

Discussion and conclusion

The *discussion* often has a relatively high proportion of marks attached to it in an assessed piece of work, because it has the potential to show your understanding of the work you have carried out. It is your opportunity to say what you think of the results, to justify the conclusions you will draw and to state any qualifications or reservations about them. It is a place where you can compare your results or solutions with those other workers have produced and comment on any differences there might be. You are expected to use logical arguments based on the work you have done to justify your conclusions.

The *conclusions* are just that, and bring the whole report to a closure. They relate to the aims or objectives as stated at the beginning of the report. If, for example, the report is about solving a problem, you would be expected to provide a solution or, if not, reasons why the solution could not be found and what might be done in the future to bring about a solution.

Here, you are not arguing a case so you are not introducing new material. It is the end of the story you are telling, although it will often have sections that identify further work that could be done or needs to be done. Conclusions often need only a short paragraph for each one, since the arguments will have been covered in the discussion.

Bibliography and references

Literature that is useful as background to the report, but has not been used specifically in the text, is listed in the *bibliography*. Long lists that have been copied from a library search or journal article are not appropriate. You should include only literature that adds to understanding and context. Remember, marks will not be given for quantity alone without quality and relevance.

References are literature *that has been cited* in the text of the report. They identify other work that has been carried out, which is specifically relevant to the topic in a way that enhances the argument being made. The purpose of references includes the identification of previous or supporting work, giving credit to other authors for quotes, theories and concepts, and providing a known authority for a particular statement or process (e.g. quoting a British Standard). References should *not* be included in the list unless they are referred to in the text.

Questions	Model Stages	Comments
What is this report about?	EXTRA STAGE	Use this question for longer reports. The answer is usually contained in an *abstract* or *synopsis*, which give a very brief description of the report including the conclusions. Sometimes, long reports have an *executive summary*, which gives more detail of the conclusions. (These items all provide a short summary of the contents and allow the reader a rapid overview of the important points.)
What is the aim and why are you doing it?	SITUATION & PROBLEM	The answer is usually contained in the *introduction*. This will indicate what is to be covered, the aims of the report and the structure to be used. A *table of contents* before the introduction will help to show the structure. (This part tells the reader what the report is for and how it intends to cover the ground.)
What has been done before?	SITUATION & PROBLEM	The answer is usually contained in a section entitled *background* or *previous work*. This will usually explain a current situation and/or describe previous work carried out by you or known to you, or information that you have researched (e.g. a literature survey). (This section gives the reader the groundwork and is a platform that the report will build on.)
What did you do?	SOLUTION	This could be called *method* or *methodology*.
What happened?	SOLUTION	This could be called *data collected* or *findings* or *results* or *design*. They might be presented as tables, graphs, diagrams, photographs, or descriptions of things. Detail might be put in the appendices so that the reader gains an overview of essential features without leaving the main text. (This section is the 'meat' of the report where the facts, design and/or processes are presented.)

Questions	Model Stages	Comments
What does the author think of what happened?	EVALUATION	This is usually called *discussion*, or less often *evaluation*. This is where the readings, findings, information are looked at to see what they mean, whether they fulfil the aims. It is the place to make comparisons, to comment on validity, possible reasons for what happened. (This is the bridge between the *solution* and the *conclusions*.) It is often done badly.
What have you found out?	EVALUATION	This is short. Usually called *conclusions* or *recommendations* or an *action plan*, or less often *implications*. It could include *recommendations for further work*. Any of these could be a separate section. No new work is introduced and the contents should logically follow from the discussion and be related to the aim or purpose. (This summarises the findings and looks ahead.)
What other literature did you get information from?	EXTRA STAGE	This acknowledges the literature used in finding out what is known already. Usually listed as *references*. There are conventions for the way these are presented (the most common is the Harvard system, see page 100).
What other books might be useful?	EXTRA STAGE	This is a list of books or *bibliography*, not necessarily directly used, but useful as general background.
What detail is important, but not necessary for the main text?	EXTRA STAGE	This is material that is not necessary for the main argument, and would be placed in *appendices*. Much of the detailed data would be here, as well as other work that an interested reader who wants more detail could refer to.

Referencing material

Formal reports require the references to be set out in a specific way. The Harvard system (or author-date system) is commonly used. The system can vary in ways such as punctuation, capitalisation, abbreviations and the use of italics. The most important principle is consistency. Different institutions and academic disciplines may prefer different systems, so check before you start writing. Referencing is necessary to avoid accusations of plagiarism (presenting somone else's work as your own), to verify quotations and to enable readers to identify and consult items in order to follow up a cited author's argument.

In the Harvard system, references are quoted by author name and date in brackets in the text of the report. The details are given in the references section at the end, listed in alphabetical order and using the following layout:

Books

- Name of author/editor
- Dates of publication, in brackets
- Book title, including other information, underlined or in italics
- Series title
- Edition (if other than the first)
- The publisher's name
- Place of publication
- Page numbers (if applicable)

Articles

- Name of authors, with initials
- Date of publication, in brackets
- Title of article in single quotation marks
- Name of journal/periodical underlined or in italics
- Volume number, issue number and page numbers

Media (e.g. video)

- Author
- Date of publication
- Title underlined or in italics
- Format and length
- Publisher
- Place of publication
- Accompanying material

Internet

- Author
- Date of publication (if you cannot establish a date, use n.d. (no date))
- Title or page, underlined or in italics
- Edition if applicable
- Type of medium if necessary
- Name and place of publisher, sponsor or host of the source
- Date viewed
- Web page or site address

Examples of references

A book

Godfrey J.C. & Slater M.J. (Eds.) (1994), *Liquid-liquid Extraction Equipment.* John Wiley & Sons, Chichester.

An excerpt or chapter from a book

Gourdon C. (1994), Population balance based modelling of solvent extraction columns. In: Godfrey J.C. & Slater, M.J. (Eds.) *Liquid-liquid Extraction Equipment.* John Wiley & Sons, Chichester, pp 137-226.

A journal article

Semple, T.C. and Worstell, J.H. (1996), 'Rethink your model for improving processes', *Chemical Engineering Progress.* 92(3), pp56-61.

A paper from a conference

Marshall, V.C. (1989), What goes wrong? In: *5th Annual European Summer School on Major Hazards: the assessment and control of risk.* 24th-28th July 1989 Christ's College, Cambridge. London, IBC Technical Services Ltd.

Other media

Cleese J. & Jay A. (2000), *Meetings Bloody Meetings*, Video recording, Video Arts, London.

Internet sources

Cross P. & Towle K. (1996), *A guide to citing Internet sources* [online], Poole, Bournemouth University.

http://www.bournemouth.ac.uk/library/using/guide_to_
citing_internet_sourc.html. Viewed 24 Jun 1997.

Referencing in the text

The author and the date are given in the text in the Harvard
system, and if necessary specific pages. It can be inserted at
the end of the sentence in brackets or integrated in the text
with the author's name, and the year directly after in brackets
(see the examples below). The reader can then refer to the
full reference at the end of the report/essay.

Examples

"Modern methods of liquid-liquid extraction are reviewed by
Godfrey & Slater (1994, pp 10-25) in their book in which a
number of authors describe different types of equipment."

"New methods of risk assessment have been proposed in
this area (Marshall)."

"In the chemical engineering field, modelling is now being
radically changed as a way of improving the processes
(Semple & Worstell)."

If an author(s) is cited more than once then the references
are given with the date.

Example

"(Marshall, 1989)" and "(Marshal, 1990)" in the text, and
then listed in the References with Marshall, 1989 before
Marshall, 1990.

If an author(s) is cited more than once within the same year
then the references are distinguished in the text and list of
references by lower-case letters, i.e. a, b, c...

Example

"(Worstell, 1996a)" and "(Worstell, 1996b)" in the text,
and then listed in the References with Worstell, 1996a listed
before Worstell, 1996b.

If there are more than two authors the reference in the text
can be the first author plus the abbreviation "*et al*" ("and
all").

> **Example**
>
> The reference "Northedge A., Thomas J., Lane A. & Peasgood A. (1997), *The Sciences Good Study Guide*, Open University Press, Milton Keynes."
>
> Can inserted in the text as: "The Open University provide excellent general study skills material for mature students (Northedge *et al*)."

The text citation for electronic sources would be the same as for books and journals using author and date. Bibliography lists are also in author alphabetical order with same layout as references.

A common mistake is for students just to list references in the back of a report without citing them in the text. It is expected that *ALL references are cited in the text* to make a point or acknowledge where data, a concept or a quote is taken from. This is both ethical and adds weight to the argument in the report. Short quotes are put in quotation marks and longer ones (above a couple of lines) are usually put in a separate paragraph.

Notice the difference

> **Exercise 6.4**
>
> Students are often unclear about the difference between:
>
> - *Discussion* and *Conclusions*
> - *References* and *Bibliography*
>
> Re-read the sections describing these on pages 96 and 97, so that you clearly understand the differences. Think about whether you make the difference clear in the reports you write.

Using graphics

These can take the form of diagrams, photographs, graphs, tables and drawings. The guidelines are similar to those for using visual aids in presentations (see page 122). They need to be easy to find in the text (either obvious or the page number given). Complex tables, which are not needed for the main argument and interrupt the text, should be put in an appendix.

The text should identify the diagrams, etc., and explain where they are and what can be discovered from viewing them. You should not just use a diagram without explanation.

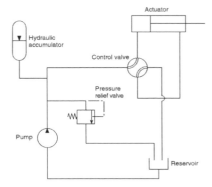

Remember, from the table on pages 94 and 95, that some of the common errors were associated with diagrams and graphs.

When using graphics, always ask the questions 'why am I including this figure?' and 'how will this add to what I am saying?'

Starting to write

In report writing, it is unusual for people to get it right first time. Most book authors go through numerous drafts, and even in shorter pieces you need to allow the time to have at least a second or third

look at it. If possible it is useful to have time in between drafts to let your ideas 'settle' and then you return to them fresh.

You should find that the actual process of writing helps you to think. The idea that you do something and then 'write it up' is not always helpful. Your thoughts and ideas change as you put them on paper and may change again when you come to read them some time later.

As you write, the physical process of constructing a logical framework develops into a thought-provoking mental exercise. This deepens your ideas, makes explicit areas you may not previously have seen clearly and leads on to new questions and avenues to explore and consider.

A good place to start writing might be a Table of Contents— in other words a structure for what you are going to say. It is hardly ever too early to start writing this. You will often need to change it later but it focuses your thinking and starts the 'flow'.

Word processors have made report writing easier because you can build on an earlier draft by cutting and pasting. This enables you, at first, just to get ideas and information into a draft without worrying about structure or English too much. This may be the solution to the common experience in writing of 'being blocked' or not being able to 'put pen (finger!) to paper (keyboard!)'.

Just starting to write can promote a creative frame of mind and could be viewed as a way of 'brainstorming'. It is a bit like 'downloading your brain' into the computer's hard disc. You can return to this later, try out different structures, improve the clarity and the English and gradually produce a logical, structured argument.

This method is often used by journalists when they are putting a story together. It means that in the first run through you do not have to worry about the mechanics of writing but merely getting down the ideas and facts. You can return to 'polish' later.

If you are stuck, you could try a 'mind map' of all your ideas and sources as described in Chapter Three. You could also use this to get down your ideas as the starting point in the process of structuring.

Another way of starting is to discuss with someone else what you are going to say. In speaking rather than writing, this can start your 'flow' and 'clarify your mind.'

Arguing a case

In short reports, the structure itself (which approximates to situation-problem-solution-evaluation) produces the argument and conclusions. In longer reports and essays, the content will be complex, and more thought has to be given to 'arguing a case'.

Think of yourself as a lawyer in a court case planning your 'summing up'. Go back over the evidence, put it in a logical way that builds to a conclusion and highlights the

Exercise 6.5

The Table of Contents can be a good place to start to write a report or essay. Imagine you are to write a report on 'The future prospects for the food-production industry in the UK over the next 10 years' (or another industry that you personally are more interested in). Make up a Table of Contents of what you think the report might contain if it were, say, 4000 words long. You could start by using the structure on pages 98 and 99, and then modifying it for your particular purpose.

important points for the jury (assessor) to look at. Structure your report so that it supports your argument.

In order to produce your arguments, you need to use both divergent thinking and convergent thinking:

- Divergent thinking is needed when you are writing the broad overview, considering a wide range of strategies, using many different types of evidence from a range of sources, discussing results in an informed and creative way or making innovative or broad-based recommendations.

It is a type of brainstorming and the ideas produced usually need narrowing down. This narrowing down requires:

- Convergent thinking, in which you bring the argument to a conclusion that relates logically to the results and discussions dealt with in earlier sections of the report. This process requires judgement in selecting the most appropriate items, a skill that is not always quantifiable.

You might find it useful to think of a report as two snapshots, one of where you were when you started and the other where you finished up (conclusions), as shown in the diagram below.

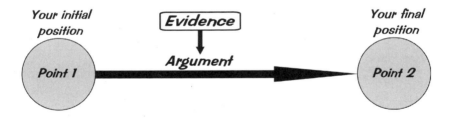

Point 1 is described by the Introduction, Background, Literature Survey, etc., and Point 2 is the Conclusion, Action Plan, Recommendations, etc. The description of how you got from point 1 to 2 is Method, Data Collection, and Discussion. The whole report is an Argument using Evidence that shows how you got from 1 to 2, and the discussion section is the major contributor to that. The argument will be the theme that runs through the whole report and keeps it together and focused. The argument was the Evidence and Logic required to reach a conclusion at point 2.

Feedback guidelines for report writing

Exercise 6.6

When next you have to write a report, use the checklist below to record and improve what you do. The checklist will take you back to many of the points raised in the text. The list is an aide-memoire to see what 'mini-skills' you are showing in a given report. Assess the extent to which you have achieved each item. Mark yourself out of ten, with 10 being 'completely achieved' and 0 being 'not achieved at all'.

You can copy the checklist to a colleague and ask for feedback on your strengths and areas for improvement.

SKILLS DEMONSTRATED:	Score out of 10
Clear objectives	
Written in a style appropriate to purpose, content and company or institution	
Argued a good case, based on the data presented	
Logical sequence	
Accurate and clearly presented data	
Good layout and logical structure	
Guidance through the structure for the reader	
Material referenced and labelled correctly	
Well-organised content	
Appropriate content	
Appropriate use of words for the readers	
Accurate spelling and grammar	
No more words than necessary for the purpose of the report	
Clear and unambiguous sentences	
Clear explanations	
Appropriate use of examples, illustrations	
Well-written conclusion, summarising findings and meeting report objectives	
All the stated objectives covered	

How are you doing?

Exercise 6.7

Use the table below to consider 'How are you doing?' in solving the problem of writing reports well. Use the questions to check this out.

QUESTIONS	NOTES AND COMMENTS
Problem What are the aims of the report? Have I defined the problem in sufficient detail? How will I know when I have achieved my aims?	
Solutions Have I considered all the solutions? Have I considered all the techniques? Have I made use of all my qualities and skills?	

Selection

Have I consciously selected the solution?
Have I made a plan of what, where, when…?
Does it solve the problem?

Evaluation

Have I achieved my aims (solved the problem)?
Could the solution be improved?
In the future I will…

Chapter 7

PROBLEM: How can I give good talks?

The general introduction given in Chapter Five can be read as an introduction to this chapter. In particular, consider the *Purpose*, *Language*, *Structure*, *Content* and *Delivery* sections and the way they might be applied to giving talks.

A talk can take many forms, from a small input to a meeting, to a presentation at a small departmental meeting, up to presenting a paper at a large international conference. It can involve using no visual aids at all, or many, with a variety of equipment such as overhead projectors, computers, video players or flip-charts.

You could use the problem-solving model to tackle the 'problem' of giving a talk. The problem involves what to communicate, the tools available (the words you use, visual aids, handouts, etc.), how to use the tools most effectively, and then checking that you have achieved the objectives. Like other forms of communication, it is useful to identify the objectives of the communication—this helps keep the talk and the preparation focused. As with writing reports, the preparation is most important, and you can use the headings of *Purpose*, *Language*, *Structure* and *Content* to do a lot of the work before you enter the room or lecture theatre.

Example

Kieran was asked to present the sales figures for all the reps in his team at a Divisional Review Meeting. The purpose was to present the figures and explain the reasons for the various trends. His target audience was a group of Divisional Managers and the Sales Director, who were all familiar with the background to his presentation. He knew, however, that the Managers were often unaware of the special circumstances in his area and he would need to summarise these. The structure of the talk could be simple:

1. A brief introduction that sets the scene and outlines the special circumstances;
2. The facts in terms of sales figures, customers, etc.—graphs or tables;
3. The reasons for the figures;
4. Any actions to be taken.

He would get feedback as he spoke by watching their faces and body language and encouraging the audience to ask questions.

Kieran knew that nervousness would be his main problem during the delivery. Before the presentation he intended to prepare well. This would give him confidence and help him to keep calm. His preparations would include the production of a good PowerPoint presentation and informative handouts. It would be important to arrive in good time for the meeting and not have a last-minute rush. He knew that he would speak clearly, but had to remember to make eye contact from time to time during the presentation.

The audience were all familiar with the subject matter and used similar language, so he was fairly confident that this would not be too much of a problem. He did, however, need to be careful about the content to keep the talk concise. These were busy people and would quickly 'switch off', so he would need to keep the talk well focused.

Exercise 7.1

Think of an example of a presentation or talk that you have attended and identify how the presenter attended to the *Purpose*, *Language*, *Structure*, *Content* and *Delivery*.

Write down the strengths and weaknesses of the example in these areas.

Common mistakes

Remember 'Sod's Law'—if anything can go wrong, it will. Keeping things simple is another way of keeping control and minimising risk. This doesn't mean you have to be boring, but experimentation should be associated with experience. By all means take a chance, but always think 'what if...?' and have a contingency plan.

Is there anything you do which can be a distraction? If so, pay some attention to it. Don't become paranoid about it, though, and remember some mannerisms are part of people's character.

Anxiety causes many people to talk to the projector screen. The audience want you to speak to them, so stop talking when changing slides or referring to visual aids or notes. Keep plenty of eye contact. If you want to refer to your slides, look at the overhead projector platen or computer monitor rather than the screen behind you.

Stand still and don't fidget. If the audience wanted to see a dance, they would go to a concert. Try to keep your feet still, and let your mouth and hands do the talking.

Exercise 7.2

A list of common mistakes is given in the table below. It includes a column where you can assess how likely you are to make these particular mistakes. Rate yourself on a scale of 0 to 10, where: 0 = I will always make this mistake and 10 = I will never make this mistake.

This will give you some idea where you could improve and where you already have useful strengths.

Mistakes	Your score out of 10
Going too fast with no pauses	
Covering too much ground	
Missing out the introduction and/or conclusion	
Assuming too much knowledge or understanding of words from the audience	
Not having clear objectives	
Not having a clear structure or not communicating the structure to the audience	
Too many slides	
No eye contact or interaction with the audience	
Disorganised content	
The words on the slides are too small or there are too many	
'Flat', monotone delivery	

Structure for presentations

Like any medium used to convey information, a talk must have a defined and logical structure. In this way, the audience is introduced to the presentation and has an idea of what areas are to be covered; these areas are then expanded and finally the talk is concluded in summary.

So it must have the following:

- Introduction   ⇨   Beginning
- Main content   ⇨   Middle
- Conclusion   ⇨   End

In other words:

- Tell them what you're going to tell them
- Tell them
- Tell them what you've told them

The structure is:

Introduction and overview
(Tell them what you're going to tell them)

Purpose and context.
Overview of main points and structure.

Identify Yourself

Your audience will want to know who you are and what role you have played in the things you are going to talk about.

State what you are going to cover

Having found out who you are and what you are going to talk about, they will probably also want to know how long you are going to talk for—is this to be a brief statement or will it be a long haul...?

Bear in mind what the audience already knows

The audience must not have their intelligence insulted by being told the obvious, nor become lost in a fog of technical jargon.

Main content
(Tell them)

The main content of the presentation will inevitably take up the bulk of the time, and should cover the topics introduced at the start of the talk.

Cover relevant areas

All of the topics must be relevant to the scope of the talk, and the time devoted should be roughly in proportion to their importance. Avoid digressions—this is most easily done by looking critically at the *content* during preparation.

Don't try to cover too much

Bear in mind the time allocated—you will find that it goes very quickly, and you can't cover as much as you expected.

As a guide, expect to spend at least a couple of minutes on each topic, and use not more than one slide for every three minutes allocated. Avoid 'flashing' slides—if it is only required for three seconds, it isn't required at all!

Each area should have a visible structure

Each topic should be like a mini presentation, with its own structure, and there must be a logical and identifiable progression from one topic to the next. This may be by putting up a new slide, a short pause, or just giving the new topic's title.

Summary and conclusions
(Tell them what you've told them)

Re-statement and review of main points.

Conclusions and implications

At the end of each section, and the end of the entire presentation, a summary is required to highlight the major points covered.

Sum up briefly and end on a high

All relevant material should have been introduced previously, and the conclusion should consist solely of a précis of this.

Leave an impression (a good one!)

The conclusion is the last part of the presentation, so try to leave a final good impression that will stick in the audience's mind after you have finished.

This structure fits the 'situation-problem-solution-evaluation' structure which works for most technical and scientific reports (see Chapter Five). You may find, however, that there are times when you need to use something different to make sense of the material. For example, there are times when telling a story in chronological order is best.

Preparation

Start early

Most poor presentations are at least partly as a result of insufficient time being allowed for the preparation. They can take quite a while to prepare, and a late start will mean that no time is left for rehearsing and polishing the material and methods used.

Decide on areas to be covered

The first decision to be made is what the presentation is to be about. This may seem obvious, but it is easy to stray from the point. The breadth and depth of the material covered determine how long it will take; where time is limited, you must decide whether to cover a small part in depth, or a wide scope but only superficially. This judgement is crucial to the effectiveness of the final result.

Prepare good presentation material

The better the material you start with, the greater the potential for an effective and interesting talk. Don't try to be complex, with either the material or any props used, and thereby minimise the potential for disasters.

Rehearse

It is immediately obvious to an audience which presentations have been well rehearsed and which have not. Rehearsal is vital, because it allows you a unique opportunity to find weaknesses, check timing and improve your style and delivery. It is also a huge confidence booster. Using a video camera is useful, because it is the only way in which you can see yourself as others see you. This can be a sobering experience!

Exercise 7.3

Make a video recording of a talk you are to give. Watch the video (you could invite some colleagues or friends to help you with this), and evaluate it critically the way the audience would. Look for the good points, and those which you could improve. Be honest with yourself, but be objective. Some people are very self-critical, which is fine as long as this is used constructively.

After the first rehearsal, you will probably find there is a substantial amount of rework to do. When this is done, rehearse again, using video, and have another look so you can fine-tune the final work.

Tip: when using a video, don't have a video operator—set it up and then leave it unattended while you rehearse. This way you can forget about it as far as possible.

Presenting yourself

Introduction

You are 'on' from the moment you enter the room or get up to walk to the front. Make 'gentle eye contact' and stand still at the beginning. Try to look at everybody in the audience a few times during your talk.

Image

The way you dress, how you walk, how 'still' you seem, will give an impression. Make sure you dress appropriately and as the audience would expect, but also in a way that is comfortable for you and makes you feel good about yourself.

Body language

Keep your head up. Take your time and don't speak too quickly. Have 'open' gestures appropriate to the space (for example, open hand movements, not crossed arms). Before starting, take time to arrange the space at the front so you are comfortable—notes in the right place, a table in a convenient place, etc. Always make sure that you know how the audio-visual equipment works before you start.

Managing nerves

This is a 'con trick'. You *will* feel nervous, but the trick is not to show it. Shoulders are important—they come up under tension. Try to relax them. Face the fear. Consciously relax and breathe a little deeper. The physical effects of anxiety are shaking, flushing, sweating. Sometimes people get clumsy or start to fidget. Try to avoid annoying habits (asking your friends for feedback!). Minimise the *er*s, *umh*s, *OK*s, etc. When you are nervous you start to be forgetful. You take short breaths as the diaphragm tenses up. Some people have a dry mouth, so make sure water will be available. With nerves, the voice increases in pitch and you speak more quickly. The best rate is about 100 words per minute; this can go up to 200 under stress. Deliberately speak more slowly and lower the pitch (slightly!) to compensate.

   Don't be distracted by all the things I have told you might go wrong! Now that you know about them they are less likely to catch you unawares.

'Stagecraft'

Actors use the stage for effect. There are similarities (and also, of course, differences) when you give a presentation. When you are standing, as is often the case when addressing a large group, this will affect the emphasis of what you say. Downstage (nearer the audience) is more intimate; upstage (away from the audience) is more inclusive of those at the back. Your movement should be driven by decision, not anxiety. Once you have moved to where you want to be, stay still. Go from 'stillness' to 'stillness'. Don't hide behind the overhead projector or a flip-chart. Control the group by where you are. Know when you want the 'spotlight' and when you want the audience to focus on a visual aid. Don't fiddle with a pointer or pen—put it down!

If something goes wrong with the equipment during your presentation, keep calm. If you are calm your audience will not feel uncomfortable. Think about the possibility of equipment problems beforehand and have some idea about what you will do if things go wrong. Try to stop talking while it is sorted out and tell your audience that you are doing this. Try not to give them a commentary on how you are trying to get it fixed!

Remember, when something goes wrong most people will feel sympathetic to your dilemma and if you handle it calmly they will still be 'with you'.

Voice

Build the dynamics of a sentence to the end: don't 'tail off'. Try to keep your voice sounding interesting rather than monotonous. It is what we do in natural conversation, but anxiety can rob us of our naturalness. We have a variety of ways of creating interest and meaning with our voice:

- *PITCH*
- *PACE*
- *PAUSE*—this can be a powerful tool; it allows you to catch your breath, lets the audience catch up, provides a space for feedback and questions
- *VOLUME*
- *TONE*—conveys feelings (e.g. hardness, softness, warmth, sarcasm)

Speak at the front of the mouth, slightly exaggerating the mouth movement. Consonants make the articulation when the teeth and lips meet. Changing the shape of the mouth makes the vowels. Don't hide your mouth with your hand or your notes. People 'hear' a great deal by 'lip reading'.

Exercise 7.4

Write down your own strengths and weaknesses concerning the personal side of *Presenting Yourself*. If you have made a video, go back and look at yourself again with these things in mind.

Write down how you will overcome some of your weaknesses. Try to be as specific as possible. Remember the SMART goals in Chapter Two.

Use of visual aids

Overhead projector or computer slides and other visual aids need careful thought. You need to plan when and how to use them and, as with all the elements of communication, consider why you need them and

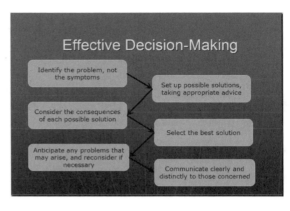

what they will contribute to your presentation.

Are they:

- to copy from?
- to add to the spoken word?
- to replace the spoken word?
- to show a picture?
- to reinforce a point?
- to show data in the form of diagrams, maps, graphs etc.?
- to demonstrate?
- to summarise?
- something else?

Always ask yourself whether they will enhance or distract from the *Purpose* of the talk.

Once you have decided on your visual aids, make sure that you give the audience time to read them. When they are on the screen they will distract the audience from what you are saying. Sometimes it is better to stop speaking for a moment and allow assimilation time. On other occasions it might be better to read through them with the audience. Think about the timing of slides and at what point they will be most effective.

When you have finished with the slides, switch off the projector or cover up the slide so the focus is back to you. You can use slides as prompts, but it is easy to be mesmerised by the screen and forget to look at the audience. It is usually better to read by looking down on the overhead projector platen or laptop computer screen, while still facing the audience. Remember to move away from the overhead projector and then start speaking. If you need to identify words or pictures on a slide, think about how you will do this—pointer, laser, computer cursor, etc. When dealing with equipment it is useful to follow the procedure of 'TOUCH—TURN—TALK':

- TOUCH—switch on the equipment and adjust it
- TURN—move away from it so that you are not tempted to 'fiddle' with it
- TALK—once away from it, start talking

Some guidelines on using slides are:

- not too many—the audience needs assimilation time and a finite time is needed for each slide (it can be useful to estimate this in advance)
- not too many words on one slide—again, assimilation time
- use colour, but not too much
- use large print (at least 22-point font) so that it can be seen
- not too many points on one slide (a maximum of five or six)
- most screens are square so do not use all of an A4 sized slide
- graphs and diagrams need to be as simple as possible so that everyone can see them—make sure the axes don't disappear!

Remember there are other visual aids you can use:

- you are one yourself!!
- things to show at the front—specimens, products
- things to be handed around (but remember they will take attention from you)
- large wall charts (but make sure they can be seen from the back)
- audio or video recordings
- 35mm slides
- demonstrations

Sometimes it might be useful to ask questions, invite the audience to ask questions or have a discussion to involve them more.

Handouts can be used in a variety of ways. For example, they can supplement your talk, give data or graphs that you want to talk about, or give a summary of main points. If you are going to use handouts, tell the audience about this in advance, so that they can decide when to take notes and will know what will be provided.

If you have been at a lecture where you have laboured to take copious notes, only to be given all the details at the end, you will know irritating this can be.

Remember if you give a handout then immediately start to talk, you will have lost your audience—they will be reading the handout!

Exercise 7.5

Write down your own the strengths and weaknesses concerning using visual aids.

Write down how you will overcome some of your weaknesses. Try to be as specific as possible.

Computer presentations

The laptop computer and programs like Microsoft PowerPoint can be very valuable and flexible tools in producing really good presentations. You do need, however, to be careful that you do not spend so much time  with the technology that you do not fulfil all the basic criteria needed in a successful communications process.

PowerPoint and similar programs will encourage you to produce clear slides, with only a few points on each. These can be used directly through a laptop computer, or printed onto suitable acetate sheets to be used as overhead projector slides if required.

High-quality graphs, pie charts, histograms, etc. can be produced directly from a computer spreadsheet program such as Microsoft Excel and printed out on an A4 sheet as handouts to go with your slides. They can be reduced in size to allow up to six to fit on an A4 sheet. If there is plenty of room left on the paper this can be a good way for the audience to take notes about the slides as you talk.

PowerPoint presentations using a data projector are now standard in many situations. This gives you the possibility of adding various features, which can build up slides and then let them fade, etc., or import Microsoft Excel tables and build them up on screen. For certain applications this can be very effective, but if overdone can be very distracting. It is also relatively easy to import photographs from a scanner or digital camera and integrate them with the text.

Use these tools only when you are confident with the technology. Remember you probably still need the slides in hard copy in case things go wrong!

Feedback guidelines

Exercise 7.6

Using the feedback guidelines below, assess one of your presentations and also ask a friend to do it for you. Compare the results and fill in what you consider to be the strengths and weaknesses of the talk.

Assess the current level of your skills for each point. Rate yourself on a scale of 0 to 10, where 0 = *I did this extremely badly* and 10 = *I did this perfectly*. They will not all be relevant for every talk you give.

SKILLS DEMONSTRATED:	Score out of 10
Managed time well	
Used visual aids well	
Showed confidence	
Established eye contact	
Appropriate body language	
Conveyed enthusiasm	
Talked to audience, not at them	
Clear communication (volume)	
Clear communication (articulation)	
Clear objectives	
Good structure	
Clear explanations	
Varied delivery to maintain interest	
Appropriate pace for material	
Appropriate use of examples, illustrations	
Adjusted to audience's needs and feedback	
Created interest in content	
Appropriate content	
Answered questions confidently	
Gave clear and accurate answers to questions	

Strengths of the presentation:

Ways I could improve:

How are you doing?

Exercise 7.7

Use the table below to consider 'How are you doing?' in solving the problem of giving talks effectively. Take a specific talk you have given and use the questions below to check this out.

QUESTIONS	NOTES AND COMMENTS
Problem What are the aims of my talk? Have I defined the problem in sufficient detail? How will I know when I have achieved my aims?	
Solutions Have I considered all the solutions? Have I considered all the techniques? Have I made use of all my qualities and skills?	
Selection Have I consciously selected the solution? Have I made a plan of what, where, when...? Does it solve the problem?	
Evaluation Have I achieved my aims (solved the problem)? Could the solution be improved? In the future I will...	

Chapter 8

Useful books

There are many useful books on study skills, and here are few that you might look at.

Of all these the one you should most consider buying is *The Student Writer's Guide: An A to Z of Writing and Language* as a reference book. It has many of those things that we often get wrong—for example when to use 'effect' and 'affect', and when to write numbers in words.

1. Allison B. & Race P. (2004), *The Student's Guide to Preparing Dissertations and Theses*, RoutledgeFalmer, ISBN 0415334861.
2. Blamires H. (2000), *The Penguin Guide to Plain English*, Penguin, ISBN 0140514309.
3. Bourner T. & Race P. (1995), *How to Win as a Part-Time Student*, Kogan Page, ISBN 0749416726.

4. Buzan T. (2003), *Use Your Head*, BBC Consumer Publications (Books), ISBN 0563488999.
5. Buzan T. (2003), *The Speed Reading Book*, BBC Consumer Publications (Books), ISBN 056348702X.
6. Fairburn G.J. & Winch C. (1996), *Reading, Writing and Reasoning*, Open University Press, ISBN 033519740X.
7. Kent N. (1990), *The Student Writer's Guide and A to Z of Writing Language*, Stanley Thornes, ISBN 074870499X.
8. Northedge A., Thomas J., Lane A. & Peasgood A. (1997), *The Sciences Good Study Guide*, Open University Press, ISBN 0749234113.
9. Powell S. (1999), *Returning to Study*, Open University Press, ISBN 0335201318.
10 Honey P. & Mumford A. (2000), *The Learning Styles Questionnaire*, Peter Honey Publications, ISBN 1902899075.

Exercise 8.1

Find a study skills book from the library or the list above and write a review of about 300 words saying why you like it, how it has helped you to approach your studies and what suggestions or ideas you will use in the future.

Appendix I

Kegworth air crash report

Photo: Mike Rifaat

In 1989 a Boeing 737 airliner, en route from London Heathrow to Belfast, suffered an engine failure and crashed onto the M1 motorway close to East Midlands airport. Like many incidents of this kind, there were numerous factors that contributed to the crash, but a major one was poor communication. The following are two extracts from the official report of the crash, reproduced with the permission of the Civil Aviation Authority. The extracts here deal with a specific communication problem, that of the information provided by the engine instruments, specifically which engine each instrument referred to. Elsewhere in the report, reference is made to some confusion amongst the crew as to which engine had failed, resulting in the 'good' engine being shut down; when the damaged engine suffered a catastrophic break-up, the aircraft was unable to reach the runway to land safely.

You may also find it useful to look at the way this part is structured, and the language and style used.

"*Synopsis*

The accident was notified to the Air Accidents Investigation branch during the evening of the 8 January 1989 and the investigation was initiated on-site at 0400 hours on the morning of the 9 January. The AAIB Investigating Team comprised Mr E J Trimble (Investigator in Charge), Mr J D Payling (Operations), Mr C G Pollard (Engineering, Powerplants), Mr S W Moss (Engineering, Systems), Mr R D G Carter (Engineering, Structures), Wing Commander D Anton, RAF Institute of Aviation Medicine (IAM) (Survivability), Mr P F Sheppard and Miss A Evans (Flight Recorders). In addition Mr R Green, Head of the Psychology Department of the RAF IAM, was co-opted to investigate the human factor aspects of this accident and Captain M Vivian of the Civil Aviation Authority (CAA) Flight Operations Department was co-opted to assist the final assessment of the operational aspects.

G-OBME left Heathrow Airport for Belfast at 1952hrs with 8 crew and 118 Passengers (including 1 infant) on board. As the aircraft was climbing through 28,300 feet the outer panel of one blade in the fan of the No 1 (left) engine detached. This gave rise to a series of compressor stalls in the No 1 engine, which resulted in airframe shuddering, ingress of smoke and fumes to the flight deck and fluctuations of the No 1 engine parameters. Believing that the No 2 engine had suffered damage, the crew throttled that engine back and subsequently shut it down. The shuddering caused by the surging of the No 1 engine ceased as soon as the No 2 engine was throttled back, which persuaded the crew that they had dealt correctly with the emergency. They then shut down the No 2 engine. The No 1 engine operated apparently normally after the initial period of severe vibration and during the subsequent descent.

The crew initiated a diversion to East Midlands Airport and received radar direction from air traffic control to position the aircraft for an instrument approach to land on runway 27. The approach continued normally, although with a high level of vibration from the No 1 engine, until an abrupt reduction of power, followed by a fire warning, occurred on this engine at a point 2.4 nm [nautical miles] from the runway. Efforts to restart the No 2 engine were not successful.

The aircraft initially struck a field adjacent to the eastern embankment of the M1 motorway and then suffered a severe impact on the sloping western embankment of the motorway.

39 passengers died in the accident and a further 8 passengers died later from their injuries. Of the other 79 occupants, 74 suffered serious injury.

The cause of the accident was that the operating crew shut down the No 2 engine after a fan blade had fractured in the No 1 engine. This engine subsequently suffered a major thrust loss due to secondary fan damage after power had been increased during the final approach to land.

The following factors contributed to the incorrect response of the flight crew:

- The combination of heavy engine vibration, noise, shuddering and an associated smell of fire was outside their training and experience.
- They reacted to the initial engine problem prematurely and in a way that was contrary to their training.
- They did not assimilate the indications on the engine instrument display before they throttled back the No 2 engine.
- As the No 2 engine was throttled back, the noise and shuddering associated with the surging of the No 1 engine ceased, persuading them that they had correctly identified the defective engine.
- They were not informed of the flames which had emanated from the No 1 engine and which had been observed by many on board, including 3 cabin attendants in the aft cabin.

31 safety recommendations were made during the course of the investigation.*"*

"Appendix 2.7

Engine instrumentation

Layout

The design of engine instrumentation on multi-engined aircraft is inevitably a matter of compromise. The conventional and ergonomically accepted layout is for all instruments associated with a particular engine to be organised in a column, and for all instruments of the same type to be organised in a row. It is, moreover, clearly preferable for each column of instruments to be associated spatially with the throttle of the appropriate engine. This is the basic layout illustrated in Figure 1 and the desirability of using such a layout for the primary engine instruments is clear. Secondary engine information is not required on the front panel of the flight deck in those aircraft with three-man crews, and the ideal layout of front panel engine instrumentation described above may thus be adopted.

If the aircraft is provided with only two crew members, however, then the secondary engine instruments must be accommodated on the front panel as well. They cannot be accommodated by extending the height of the columns since panel height precludes such an option if the instruments are to be large enough to remain legible.

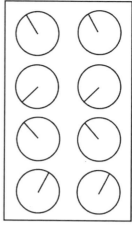

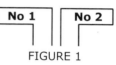

FIGURE 1

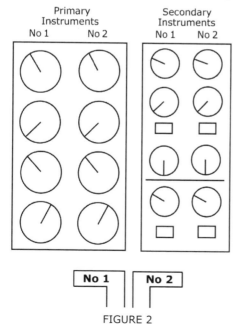

FIGURE 2

If the instruments are all to be located on the front panel, two possibilities are apparent. The first is to mount the secondary instruments to one side of the primary instruments as in Figure 2.

The second is to split the secondary instruments and mount them outboard of their respective primary instruments, as in Figure 3.

The advantage of the layout in Figure 3 is that the instruments for a given engine are all mounted together and are, if not spatially, at least cognitively, aligned with their associated power levers. This is achieved at the price of splitting the secondary instruments apart, with the associated possibility of disparate secondary readings going undetected.

Figure 2 achieves the goal of keeping the instruments paired together, and thus maximises the chances of disparate readings being detected, but does so at the price of splitting up the instruments associated with a given engine, and of losing the advantage of having all instruments cognitively aligned with their corresponding throttle levers.

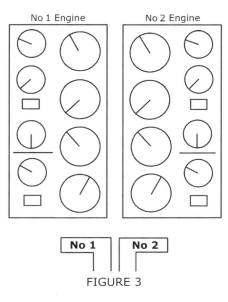

FIGURE 3

Thus, Figure 3 could fairly be judged to maximise the probability that a given failure will be correctly identified by the crew as belonging to a given engine, at the possible cost of less efficient error detection on the secondary instruments, whereas Figure 2 may be judged as maximising the probability that disparate readings will be detected at the cost of degrading the probability that this detected failure will be associated by the crew with the correct engine.

The design of the EIS

The layout of the EIS in the Boeing 737 Series 300/400 conforms to Figure 2, which has been widely used without apparent difficulty in many two-engined, two-pilot aircraft. The illumination of the display, however, might aggravate the problem of perceived misalignment of the instruments with their respective throttles. On the hybrid instruments (LED counters with electromechanical pointers) fitted to other aircraft of this type, the faces of the instruments needed to be lit from in front to show the pointers, dials and scale marks. Such lighting does not, of course, illuminate only the legends and pointers on the instruments but also the general structure and limits of the display, so that the instruments could be argued to be viewed within a structured visual frame. In the

EIS display, all symbology is edge-lit and set against a heavily contrasting background which, in an aircraft at night will be, to all intents and purposes, black. This may have the effect of enhancing the extent to which the instruments are seen as a single display rather than as two separate displays, and may degrade the extent to which deviant readings in, say, columns 1 and 3 of the matrix could readily be associated with the No 1 engine.

The next most obvious and important change made between the hybrid system and the EIS is that the full-radius mechanical pointers have been changed to short LED pointers moving round the outsides of their scales. The mechanical pointers were relatively large, white and clearly linear devices, and their orientation on the display was immediately apparent. Not only was the absolute orientation of each pointer apparent but (and perhaps more importantly) it was readily apparent whether the pointers of each pair of instruments were parallel with one another. The pointers on the LED display are much shorter than the mechanical pointers, they are the same colour as the LED counters and they move in steps. They are much less conspicuous than the mechanical pointers, acting more as scale markers, and providing less immediate directional information. They are thus less well able to give the comparative information provided by the strong cue of parallelism of the mechanical pointers. This comparative information can be obtained with certainty only by interrogating each instrument to see if the LED pointers of each pair are at the same points on the scale or by comparing the readings of the pairs of counters.

Evaluation and testing

The entire function of any display on a flight deck is to transfer information from the aircraft to the pilot, and to do so in the way that will cause the pilot least workload and will be least likely to be interpreted wrongly. Although some principles, such as those discussed above, guide the design of displays, the only way of evaluating the adequacy of a display is by experiment and trial. It is therefore important that, before any display is put into service, it is subjected not just to some form of acceptability judgement by company pilots, but to a structured assessment using average line pilots. Indeed, it could he argued that such assessments should be conducted using the least able pilots who are ever likely to use the display.

A display similar to the EIS was developed by Smith's Industries for use on the McDonnell Douglas MD88. It was held to differ from an earlier display which employed mechanical pointers, in that the colour coding of some dials was changed. The new display was evaluated by pilots employed by McDonnell Douglas and the Federal Aviation Administration (FAA). The evaluation was held to show that the new display provided clearly readable and interpretable information to the flight crew, showed whether the current state of power plant operation was normal or abnormal, indicated the engine maximum/minimum safe operating range and showed whether the system(s) operation was being accomplished in a safe manner. These results were used by McDonnell Douglas to demonstrate to the FAA the acceptability of the new display as an equivalent means of compliance with current airworthiness regulations.

The EIS for the Boeing 737 was designed to represent a minimum change from the previous hybrid display and, accordingly, it was type certified by both the FAA and the CAA as fit for its purpose. The counters remained identical in size and colour but the dials of all instruments were reduced in size. The pointers were reduced in length by approximately two-thirds and placed on the outsides of the dials but the circumference swept by the needle tips (i.e. the instrument 'size') remained the same. The EIS display was deemed to have sufficient commonality with the hybrid display to circumvent the need for pilots to be separately rated for EIS-equipped models. It was tested for proper operation, compatibility and freedom from electrical interference but it was not evaluated for its efficiency in imparting information to pilots.

Although the desire for commonality is understandable, because a number of other factors were changed between the hybrid and the EIS displays, the apparent benefit of keeping size constant may have been offset or even negated by varying others such as illumination, contrast and pointer size. The desire to maintain consistency of display format while introducing new technology was responsible for the reduction in pointer size and conspicuity, and exemplifies a general problem. LED and CRT displays possess potential advantages over old technology instrumentation that may be exploited only if the display is designed afresh to exploit them. If a new-technology display is designed simply to mimic the appearance of its precursors it may well fall into what is sometimes referred to as the 'electric horse' trap; the strengths of the old system are discarded because they cannot be duplicated, and the potential strengths of the new system are not exploited. Full-length pointers cannot be represented on the LED system because the packing density of central LEDs cannot be achieved, and because symbology cannot be over-laid, and a potentially less satisfactory pointer is substituted.

It is reiterated that the general effectiveness of any new display may be judged only by trial and experiment, but even then some criterion of acceptability must be adopted. An obvious criterion in the case of engine instrumentation is that the new display should not prove less satisfactory to those pilots who use it than the display it replaces. When the EIS was introduced for use on the Boeing 737 no such tests were carried out.

Conclusions

Although there seems to be no question that the EIS display on the Boeing 737 provides accurate and reliable informa-tion to the crew, the overall layout of the displays, and the detailed implications of small LED pointers rather than the larger mechanical ones, and of edge-lit rather than reflective symbology, do appear to require further consideration. These factors should not be ignored and the suitability of such new displays for use by airline pilots should be evaluated before they are brought into use."

Extracts from UK Air Accidents Investigation Branch (AAIB) Report No EW/C1095), *Report on the Accident to Boeing 737-400 G-OBME near Kegworth, Leicestershire on 8 January 1989,* AAIB.

Appendix II

Meetings, agendas and minutes

This appendix describes the roles of the chairperson and the secretary, both of whom are crucial to the successful outcome of most meetings. It also provides examples of an agenda and minutes, on which you could base your own paperwork.

Role of the chairperson

What is most important is that the meeting is well prepared for (especially, but not solely, by the chairperson) and well managed. An effectively run meeting just appears to happen, but in reality running a meeting is hard work and requires a lot of skill. The duties of the chairperson include:

1. to conduct the business, stick to the agenda and time-scale
2. to guide and progress the discussion towards achieving the objective of the meeting
3. to ensure everyone is heard and has opportunity to express himself/herself
4. to ensure that full discussion comes before solutions
5. to distinguish between facts and opinion
6. to ensure a decision is made, understood and recorded (this is very important especially when there has been a lot of discussion)
7. to conduct a vote, if necessary, and to ensure everyone knows the issue being voted on (it is surprising how often this does not happen)

In more-formal meetings, members address all remarks to the chairperson to avoid the personal element of discussion. This can be useful, but in student project meetings and most small meetings this would be artificial and not help the 'team atmosphere' to develop. In more-formal meetings, other devices for clarity are used such as motion, resolution, amendment, point of order. These are unnecessary for small meetings and are often better avoided.

Voting can be used with the chairperson as the casting vote but in many situations this is unnecessary and a much better 'feel' can be obtained by a general consensus. In a consensus there is agreement on the way forward while recognising that some members may have reservations. It is, however, recognised by all that, given absolute agreement is rarely possible, it is overall the best way forward, based on the views of the group.

In student groups it is often possible to take turns at the roles of chairperson and secretary, although you do miss the co-ordinating function between meetings of an overall chairperson.

Role of the secretary

The secretary can often play an important clarifying role while recording decisions in the meeting, e.g. *what shall I actually write down?* Minutes should be written up as soon after the meeting as possible—it is surprising how much is forgotten a few days after, even with good notes.

The duties of the secretary are:

1. to record who is present and the full decisions made (including what was decided, when and by whom will it be done/actioned)
2. to distribute the minutes to members
3. to arrange the meeting and get invitations and paperwork out

Note: It is not usually a good idea to try to combine the roles of chairperson and secretary—both members have much to do, and would not be able to perform effectively in a combined role. The chairperson, especially, must be able to devote all his or her attention to the proceedings of the meeting.

Meeting practice

There are many unwritten rules and standard practices that are observed (or not) in meetings. Most exist for sound, practical reasons, but some are more tradition than anything else. The following are just some guidelines, and there are many more.

1. Actions should be agreed, not placed—an action (i.e. a request for someone to do something outside the

meeting) should be agreed with the actionee, then recorded in the minutes (see below). Actions should never be changed once minuted. If the situation changes and the action required is no longer relevant, the original action should be minuted as cancelled, and a new action agreed.

2. The chairperson is in charge of the meeting, and should be respected as such. However, his or her role is not to take charge of decisions, but just to ensure they are reached. The chairperson normally does not vote, but exercises a casting vote in the event of a stalemate.

3. Items discussed at a meeting should be clearly under-stood by all the members so that the decisions taken are well informed. Items to be discussed should be supported by a structured framework, of which the agenda and ancillary papers are an important part. (The way the meeting is conducted on the day and the clear and fair guidance of a good chairperson will also play a vital role in a successful meeting.)

The structure should include the following elements:

- relevant facts, on which to make a decision
- opinions or possible solutions that are based on the facts
- decisions that arise from the solutions suggested
- a review of the process to see that everything known has been considered and all opinions and suggestions heard
- a record of all decisions, with notes of who will take what action and by when

Agendas

Agendas define the purpose of the meeting, its structure and (if necessary) proposals to be discussed which can be considered before the meeting.

All student team projects will benefit from an agenda even if it is quite short and simple. Although all meetings are different, there is nevertheless a standard format for agendas (and hence the order of business during the meeting itself). The agenda may seem trivial, and a document that can be put together easily, but an effective agenda requires a degree of planning and forethought if a logical sequence is to be achieved.

The importance of this logical sequence should be stressed— for a superb example of what can happen if matters are not logically ordered, see the video *Meetings Bloody Meetings* mentioned on page 59.

Example

To: Joe Leigh, Claire Roberts, John Cheshire, Thomas Brown, Ann Hodges (chairperson)

From: Ian Lang (secretary)

Bifurcated Rivet Progress Meeting

The seventeenth Bifurcated Rivet Progress Meeting will be held in Conference Room No. 2 from 14.00 hrs on Thursday 14 December 2005.

AGENDA

1. Apologies for absence
2. Minutes of previous meeting
3. Matters arising from minutes (not covered in the agenda)
4. Progress reports from individuals:
   4.1 Design (JL)
   4.2 Data Analysis (CR)
   4.3 Production (JC)
   4.4 Finance (TB)
   4.5 Project Management (AH)
5. Conclusions and further work needed
6. Any other business
7. Date of next meeting

The example shows a typical agenda for a progress meeting, which can be easily adapted to cover other types of meeting.

Looking at the above in detail:

Apologies. It is usual to make a list of those present—often by passing a piece of paper round for people to sign or write their name. Apologies for absence are then taken—messages sent via those present. Apologies are not assumed from those who are merely absent; it is a courtesy to send apologies if a meeting is to be missed.

Minutes of previous meeting. These should have been distributed well before the meeting, to give people time to attend to their actions (more later) and to allow them to check the minutes. The purpose of the discussion of the minutes is that all those present agree that the minutes represent a true and accurate record of the meeting. Note that comments or discussion not related to errors or omissions should be avoided at this stage. Once any corrections have been taken, the minutes are accepted by the meeting, and a file copy is often signed by the chairperson.

Matters arising. In this section, any items that arise (with the exception of those specifically addressed in the rest of the agenda) will be discussed, including progress on any actions. This may form a very small part of any meeting, or the majority of it. For this reason, it is helpful to have numbered actions to help identify each one.

Progress from individuals. This area is really specific to a progress meeting, and should be replaced by an equivalent if the meeting serves another function. For a progress meeting, avoid straying into technical areas except where necessary for the progress of the meeting. Ensure that each participant speaks in turn, and try to avoid interruptions and diversions, without unduly restricting relevant discussion.

Conclusions. It can be very useful for the chairperson to summarise the main points, and perhaps the new actions, at the end of the meeting, but he or she should wait until the very end. In this case, however, a summary of the major progress appears on the agenda, before AOB.

Any other business (often abbreviated to AOB). This is placed almost at the end of the agenda, and allows any other relevant discussion not otherwise called up on the agenda. Any points introduced during the meeting that are not directly related to the agenda item under discussion should be postponed until AOB. Participants may also introduce their own items in this section, provided they are relevant to the overall aims of the meeting. AOB may comprise no discussion, or may make up the majority of the meeting—don't assume that the meeting is over when AOB is reached!

AOB can sometimes get out of hand and prolong a meeting because of all the unscheduled topics. They are only normally allowed at the discretion of the chairperson, and in some meetings need prior notice.

Date of next meeting. Before the meeting is declared closed, a date, time and location should be set for the next meeting (if appropriate). If this is not yet known, those present should be advised as to how and when they will be informed.

In more-complex meetings the agenda could have numbered agenda papers for members to read beforehand.

Minutes

The purpose of the minutes is to record the business of the meeting. For many meetings the recording of decisions is sufficient without the discussion. The minutes are a tool for progress to be made, and decisions to be recorded and implemented. If there is a series of meetings, minutes will be needed at the next meeting. It is easy to think everyone will remember, but they don't. A decision could be many things such as an agreed principle, a planning document or an action. It should include what will happen next, with dates and briefly how it will be implemented (with people next to actions). If you put 'all' against an action, this usually means no-one will do it!

At a first meeting of a project a calendar of events (a Gantt chart, or some other form of plan) should be produced. At subsequent meetings, the progress of the project can be checked against the prediction and the plan changed and/or corrective action taken.

Example

(note: only samples of the topics covered are shown here)

ACME Engineering PLC

Minutes of the 17th Bifurcated Rivet Progress Meeting
Held on 14 December 2005

Distribution:
  As below, plus J Smith, Engineering Director

Present:
  A Hodges (Chairperson), I Lang (Secretary), J Leigh, C Roberts, J Cheshire

1. Apologies:
  T Brown

2. Minutes of previous meeting
  Page 5—JL stated that part no 35474-03 will be delayed, not 35474-01 as written. The minutes were otherwise agreed by the meeting as being a true and accurate record.

3. Matters arising from last meeting
  -

  -

-

-

Action 16.4 (CR). Obtain revised estimate for delivery of new mainframe software. CR stated that it has not been possible to obtain a delivery date until our contact returns from the USA later this week. As soon as he returns, CR will make sure he gets a definite date and report back to the next meeting.

Action 16.4—continues.

-

-

-

4.3 Production
... Otherwise, the only outstanding problems are with the large bifurcating press, which needs continual adjustment to maintain tolerances. This is increasing the cost and causing delays. AH suggested it may be worthwhile to refurbish the press. JC agreed to look into cost and timescale for this and report back to the next meeting.

New Action 17.5 (JC). Look into cost and timescale of refurbishing large bifurcating press, and report back to next meeting.

-

-

-

6. Any Other Business
CR suggested that Mike Gannon should be invited to the next meeting, since electrical plant requirements may need to increase substantially with the new machines.

New Action 17.9 (AH). Include MG in next meeting, and circulate minutes of today's meeting to him by 16th December.

JC stated that all present should be aware that the new product brochure is due shortly.

7. Next Meeting
The next meeting is scheduled for 8 January 2006.

Meeting closed at 15.40.

Index

I

J

K

L

M

Notes

Notes

Notes

Notes